**PROPERTY OF
ERAU PRESCOTT
LIBRARY**

Springer Series in
MATERIALS SCIENCE

Editors: R. Hull   R. M. Osgood, Jr.   H. Sakaki   A. Zunger

The Springer Series in Materials Science covers the complete spectrum of materials physics, including fundamental principles, physical properties, materials theory and design. Recognizing the increasing importance of materials science in future device technologies, the book titles in this series reflect the state-of-the-art in understanding and controlling the structure and properties of all important classes of materials.

27 **Physics of New Materials**
Editor: F. E. Fujita   2nd Edition

28 **Laser Ablation**
Principles and Applications
Editor: J. C. Miller

29 **Elements of Rapid Solidification**
Fundamentals and Applications
Editor: M. A. Otooni

30 **Process Technology
for Semiconductor Lasers**
Crystal Growth and Microprocesses
By K. Iga and S. Kinoshita

31 **Nanostructures and Quantum Effects**
By H. Sakaki and H. Noge

32 **Nitride Semiconductors and Devices**
By H. Morkoç

33 **Supercarbon**
Synthesis, Properties and Applications
Editors: S. Yoshimura and R. P. H. Chang

34 **Computational Materials Design**
Editor: T. Saito

35 **Macromolecular Science
and Engineering**
New Aspects
Editor: Y. Tanabe

36 **Ceramics**
Mechanical Properties, Failure
Behaviour, Materials Selection
By D. Munz and T. Fett

37 **Technology and Applications
of Amorphous Silicon**
Editor: R. A. Street

38 **Fullerene Polymers
and Fullerene Polymer Composites**
Editors: P. C. Eklund and A. M. Rao

39 **Semiconducting Silicides**
Editor: V.E. Borisenko

40 **Reference Materials
in Analytical Chemistry**
A Guide for Selection and Use
Editor: A. Zschunke

41 **Organic Electronic Materials**
Conjugated Polymers and Low
Molecular Weight Organic Solids
Editors: R. Farchioni and G. Grosso

42 **Raman Scattering in Materials Science**
Editors: W. H. Weber and R. Merlin

43 **The Atomistic Nature of Crystal Growth**
By B. Mutaftschiev

44 **Thermodynamic Basis of Crystal Growth**
P–T–X Phase Equilibrium
and Nonstoichiometry
By J.H. Greenberg

45 **Thermoelectrics**
Basic Principles
and New Materials Developments
By G.S. Nolas, J. Sharp, and H.J. Goldsmid

46 **Fundamental Aspects
of Silicon Oxidation**
Editor: Y. J. Chabal

47 **Disorder and Order in Strongly
Non-Stoichiometric Compounds**
Transition Metal Carbides, Nitrides
and Oxides
By A.I. Gusev, A.A. Rempel,
and A.J. Magerl

48 **The Glass Transition**
Relaxation Dynamics
in Liquids and Disordered Materials
By E. Donth

Series homepage – http://www.springer.de/phys/books/ssms/

Volumes 1–26 are listed at the end of the book.

Springer Series in
MATERIALS SCIENCE

Springer
*Berlin
Heidelberg
New York
Barcelona
Hong Kong
London
Milan
Paris
Singapore
Tokyo*

Physics and Astronomy

http://www.springe

G.S. Nolas   J. Sharp   H.J. Goldsmid

Thermoelectrics

Basic Principles
and New Materials Developments

With 136 Figures

 Springer

Dr. George S. Nolas
Dr. Jeffrey Sharp
Marlow Industries, Inc., 10451 Vista Park Road, Dallas, TX 75238, USA

Prof. em. H. Julian Goldsmid
University of Tasmania, 40 Osborne Esplanade, Kingston Beach, Tasmania 7050, Australia

Series Editors:

Prof. Alex Zunger
NREL
National Renewable Energy Laboratory
1617 Cole Boulevard
Golden Colorado 80401-3393, USA

Prof. Robert Hull
University of Virginia
Dept. of Materials Science and Engineering
Thornton Hall
Charlottesville, VA 22903-2442, USA

Prof. R. M. Osgood, Jr.
Microelectronics Science Laboratory
Department of Electrical Engineering
Columbia University
Seeley W. Mudd Building
New York, NY 10027, USA

Prof. H. Sakaki
Institute of Industrial Science
University of Tokyo
7-22-1 Roppongi, Minato-ku
Tokyo 106, Japan

ISSN 0933-033x

ISBN 3-540-41245-X Springer-Verlag Berlin Heidelberg New York

Library of Congress Cataloging-in-Publication Data

Nolas, G.S. (George S.) 1962-
Thermoelectrics : basic principles and new materials developments / G.S. Nolas ; J. Sharp ; H.J. Goldsmid. –
p. cm. – (Springer series in materials science ; v. 45) – Includes bibliographical references and index.
ISBN 354041235X (alk. paper)
1. Thermoelectricity. 2. Thermoelectric apparatus and appliances. 3. Thermoelectric materials. 4. Thermoelectric cooling. I. Sharp, J. (Jeffrey) 1964- II. Goldsmid, H.J. III. Title. IV. Series. – TK2950 .N65 2001
537.6'5–dc21   00-052671

This work is subject to copyright. All rights are reserved, whether the whole or part of the material is concerned, specifically the rights of translation, reprinting, reuse of illustrations, recitation, broadcasting, reproduction on microfilm or in any other way, and storage in data banks. Duplication of this publication or parts thereof is permitted only under the provisions of the German Copyright Law of September 9, 1965, in its current version, and permission for use must always be obtained from Springer-Verlag. Violations are liable for prosecution under the German Copyright Law.

Springer-Verlag Berlin Heidelberg New York
a member of BertelsmannSpringer Science+Business Media GmbH

http://www.springer.de

© Springer-Verlag Berlin Heidelberg 2001
Printed in Germany

The use of general descriptive names, registered names, trademarks, etc. in this publication does not imply, even in the absence of a specific statement, that such names are exempt from the relevant protective laws and regulations and therefore free for general use.

Typesetting: Camera-ready copy from the authors
Cover concept: eStudio Calamar Steinen
Cover production: *design & production* GmbH, Heidelberg

Printed on acid-free paper   SPIN: 10741705   57/3020/cu   5 4 3 2 1 0

Preface

The field of thermoelectrics has grown dramatically in recent years. In particular new and novel materials research has been undertaken and device applications have increased. In spite of this resurgence of interest there are very few books available that outline the basic concepts in this field. Thus it is necessary that a book be written that encompasses the basic theory and introduces some of the resent research into improved materials for solid-state cooling and power generation.

Therefore the aim of this book is threefold. First, to present the basic theory of thermoelectricity. Both theoretical concepts and experimental aspects of the field of solid-state cooling and power generation are discussed. Second, to bridge the gap between theory and application. To this end, the techniques for producing good thermoelectric materials and module design issues are reviewed. Third, to present some of the research into new and novel materials that has drawn the attention of the scientific community. The book is intended as a reference to experimentalists working in the field; however, it will also prove useful to scientists coming into the field from other areas of research. It can also serve as a useful text for graduate students.

The book is arranged in two parts. The first part (Chaps. 1–5) is devoted to the basic concepts of the field of thermoelectrics with an overview of current thermoelectric materials and devices. It contains well-established principles covering the applications of both the Peltier and Ettingshausen effects. It contains many of the salient features of the well-known text "Electronic Refrigeration" by one of the authors, H. Julian Goldsmid, but with updated information. The second part of this text (Chaps. 6–9) is a general overview of the new materials research that has been undertaken in the past decade. The choice of topics treated in this part contains much of the research that has demanded most attention in the field. It is also partly determined by the research interests of the authors.

The authors are grateful to Springer-Verlag, and Dr. Claus Ascheron in particular, for giving us the opportunity to prepare this book. It could not have been written without the support of Dr. Hylan B. Lyon, Jr. and Mr. Raymond Marlow of Marlow Industries, Inc. Their foresight and efforts directly resulted in the rejuvenation of the field of thermoelectrics in the 1990s. The recent research presented in the second part of this book would not have been undertaken without the support of the Department of Defense research programs, including the Army

Research Laboratory, Army Research Office, Office of Naval Research and the Defense Applied Research Projects Agency. The authors thank Ms. Gail Shanovich and Mr. Emmanuel Relevo for their work on the manuscript and Mr. Roy T. Littleton, IV, Dr. Terry M. Tritt, Dr. Mercouri G. Kanatsidis, and Dr. Dave C. Johnson for their contributions to the second part of the book.

Dallas, April 2001 *George S. Nolas*

Table of Contents

1 Historical Development

    1.1 Introduction 1
    1.2 Thermoelectric and Thermomagnetic Phenomena.................... 2
    1.3 Peltier Cooling and the Thermoelectric Figure of Merit 8
    1.4 Efficiency of Thermoelectric Power Generation...................... 12
    1.5 Ettingshausen Cooling and the Thermomagnetic Figure of Merit .. 13

2 Transport of Heat and Electricity in Solids

    2.1 Crystalline Solids 15
    2.2 Heat Conduction by the Lattice.. 18
    2.3 Band Theory of Solids 28
    2.4 Electron Transport in a Zero Magnetic Field......................... 36
    2.5 Effect of a Magnetic Field 44
    2.6 Nonparabolic Bands 51
    2.7 Phonon Drag 55

3 Selection and Optimization Criteria

    3.1 Selection Criteria for Thermoelectric Materials...................... 59
    3.2 Influence of Carrier Concentration on the Properties
          of Semiconductors 59
    3.3 Optimization of Electronic Properties 71
    3.4 Minimizing Thermal Conductivity..................................... 76
    3.5 Anisotropic Thermoelements 84
    3.6 Thermoelectric Cooling at Very Low Temperatures 87

4 Measurement and Characterization

    4.1 Electrical Conductivity 91
    4.2 Seebeck Coefficient 93
    4.3 Thermal Conductivity 95
    4.4 Figure of Merit... 99
    4.5 Thermogalvanomagnetic Effects.. 105

5 Review of Established Materials and Devices

5.1 Group V_2–VI_3 111
5.2 Elements of Group V and Their Alloys 131
5.3 Materials for Thermoelectric Generators 146
5.4 Production of Materials 151
5.5 Design of Modules 163

6 The Phonon–Glass Electron-Crystal Approach to Thermoelectric Materials Research

6.1 Requirements for Good Thermoelectric Materials and the PGEC Approach 177
6.2 The Skutterudite Material System 178
6.3 Clathrate Compounds 191

7 Complex Chalcogenide Structures

7.1 Introduction 209
7.2 New Materials with Potential for Thermoelectric Applications 210
7.3 Pentatelluride Compounds 220
7.4 Tl_2SnTe_5 and Tl_2GeTe_5 229

8 Low-Dimensional Thermoelectric Materials

8.1 Fine-Grained Si–Ge and Thin-Film Bi 235
8.2 Survey of Size Effects 236
8.3 Experimental Structures 242
8.4 Practical Considerations 251
8.5 Summary 254

9 Thermionic Refrigeration

9.1 The Vacuum Diode 255
9.2 Solid-State Thermionic Devices 263

References 271

Subject Index 287

1. Historical Development

1.1 Introduction

Beginning around 1990, a combination of factors – notably environmental concerns regarding refrigerant fluids and interest in cooling electronics – led to renewed activity in the science and technology of alternative refrigeration. Thermoelectric cooling is the most well established of these technologies. Closely related cooling mechanisms, thermomagnetic effects and thermionic emission, are less well established but may soon have their day. The common theme is the presence of coupled electrical and thermal currents, so that it is also possible to use the effects to generate power from a temperature differential. These basic mechanisms and the fundamentals of their implementation comprise our subject.

The physics and materials science of thermoelectrics developed mainly during two periods of strong activity. In the three decades from 1821 to 1851, the basic effects were discovered and understood macroscopically, and their applicability to thermometry, power generation, and refrigeration was recognized. The sole lasting contribution in the next 80 years was Altenkirch's derivation of thermoelectric efficiency in 1911. Then in the late 1930s, there began 20 years of progress (Fig. 1.1) that led to a microscopic understanding of thermoelectricity and the development of

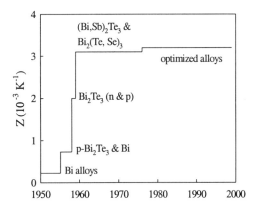

Fig. 1.1. Progress in the figure of merit of thermoelectric materials near room temperature

today's materials. The momentum from that breakthrough carried the field for most of a decade, but activity had waned by 1970.

Time will tell if the current resurgence of research will yield a breakthrough that will substantially alter the practice of electronic refrigeration. Already, though, it is clear that this research bears the stamp of its era, as was the case with the 1820–50 and 1940–60 work (Fig. 1.2). A flurry of activity showing a connection between electricity and magnetism surrounded Seebeck's discovery of thermoelectric voltage, and he erroneously incorporated magnetism into his explanation of the phenomenon. From the work of Carnot and Clausius, Kelvin deduced that the reversible heat flow discovered by Peltier must have an "entropy" associated with it. He was able to show that the coefficient discovered by Seebeck was a measure of the entropy associated with electric current. A further understanding of thermoelectric materials awaited the development of quantum theory and its application to the electronic and thermal properties of semiconductors. Taking advantage of that groundwork, thermoelectric research teams focused on the physics and materials science of the semiconductors known at that time. As for the current work, there are two unifying themes drawn from physics and materials science: the discovery of new materials with greater chemical complexity and fabrication of known materials in submicron formats.

1.2 Thermoelectric and Thermomagnetic Phenomena

1.2.1 Thermoelectric Coefficients

The basic thermoelectric circuit is shown in Fig. 1.3. Its behavior depends in part on the Seebeck, Peltier, and Thomson coefficients α, Π, and τ, respectively, which we now define. Two different conductors, a and b, have junctions at W and X. If a temperature difference is created between W and X, a voltage difference (V) appears between the two b segments. Under open circuit conditions, the Seebeck coefficient is defined as

$$\alpha_{ab} = \frac{dV}{dT}. \tag{1.1}$$

If W is hotter than X, a thermocouple ab that would drive a clockwise current is said to have a positive α. By contrast, if an imposed clockwise current (I) liberates heat at W and absorbs heat at X, then thermocouple ab has a negative Π. The rate of heat exchange at the junctions is

$$Q = \Pi_{ab} I. \tag{1.2}$$

1.2 Thermoelectric and Thermomagnetic Phenomena   3

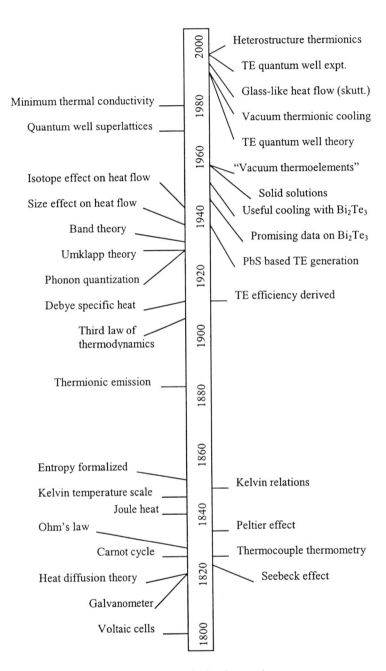

Fig. 1.2. Timeline of thermoelectric and related research

1. Historical Development

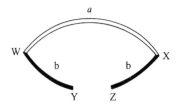

Fig. 1.3. Basic thermocouple

If current is flowing and there is a temperature gradient, there is also heat generation or absorption within each segment of the thermocouple because α is temperature-dependent. The gradient of the heat flux is given by

$$\frac{dQ}{ds} = \tau I \frac{dT}{ds}. \tag{1.3}$$

where s is a spatial coordinate.

It is useful that both τ and Π can be obtained from α, which is easily measured. Experiments have confirmed the relationships derived by Kelvin:

$$\tau_a - \tau_b = T \frac{d\alpha_{ab}}{dT}, \tag{1.4}$$

and

$$\Pi_{ab} = \alpha_{ab} T. \tag{1.5}$$

The last equation provides a fundamental link between thermoelectric cooling (Π) and thermoelectric power generation (α).

Thermoelectric cooling and power generation require joining two different materials. Therefore, it is Π and α of couples that matter in practice. However, the absence of α for superconductors has made it possible to define an absolute α and Π for individual materials. The α value for Pb–Nb$_3$Sn couples measured up to the critical temperature of the latter gave α_{Pb} for T < 18 K. Then, measurement of τ from 18 K to high temperatures [1.1, 1.2] yielded an absolute α for Pb, which became a reference material. The absolute thermoelectric coefficients also obey the Kelvin relationships:

$$\tau = T \frac{d\alpha}{dT}, \tag{1.6}$$

and

$$\Pi = \alpha T. \tag{1.7}$$

1.2.2 Electrical Resistivity and Thermal Conductivity

Two irreversible processes, thermal conduction and Joule heating, lower the performance of thermoelectric devices to less than the thermodynamic limit. A good thermoelectric material must combine a large α with low electrical resistivity ρ and low thermal conductivity λ. Further, because Joule heating is proportional to the square of the electric current and the Peltier effect is only linear in current, one cannot increase the temperature gradient indefinitely simply by increasing the current. This will be shown quantitatively in a subsequent section.

The value of ρ is defined as the ratio of the electric field $\mathcal{E}$ to the parallel current density i in the same direction and in the absence of a thermal gradient. (If the Seebeck coefficient produces an electric current, this, in turn, leads to a Peltier effect that opposes the applied temperature gradient. This situation applies to thermoelectric power generation.) In an isotropic conductor, the electrical conductivity σ is the reciprocal of ρ, but in anisotropic materials, for which the $\mathcal{E}$ and i vectors may not be aligned, $\sigma \neq 1/\rho$. The application of a magnetic field in one direction makes all materials anisotropic to some extent.

In any material, a temperature gradient leads to an irreversible flow of heat that opposes the gradient. If w is the heat conduction per unit area, the thermal conductivity is defined as $\lambda = -w\,(dT/ds)^{-1}$ in the absence of electric current. (If there is an electric current in the same direction due to the Seebeck effect, the Peltier effect opposes the applied temperature gradient. This situation applies to thermoelectric power generation.) The same remarks about anisotropy apply to thermal conductivity as to electrical conductivity. If a substantial portion of heat conduction is electronic, again a magnetic field may induce anisotropy where there was none.

1.2.3 Thermomagnetic Coefficients

Application of a magnetic field changes the coefficients already defined and leads to new effects as well. We will deal with a simple case in which an isotropic sample is subjected to a magnetic field (along the z axis) that is transverse to the current flow or temperature gradient (along the x axis) established in the absence of the field. In this situation, transport in the x direction is described by longitudinal thermogalvanomagnetic coefficients. The new effects appear transverse to the x axis.

There are four transverse effects, two in the form of an electric field along the y axis, and two in the form of a temperature gradient along the y axis. Figure 1.4 shows the sign conventions for all of the transverse coefficients. If there is a current

1. Historical Development

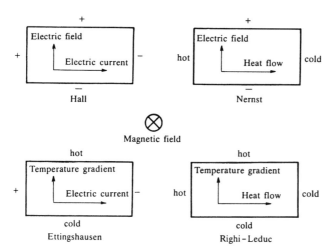

Fig. 1.4. Transverse thermogalvanomagnetic effects. The coefficients are positive if the effects are in the directions shown in the diagram

density i_x and a magnetic field B_z, a transverse electric field $\mathcal{E}_y$ appears that is proportional to both. This is the Hall effect. The Hall coefficient,

$$|R_H| = \frac{\mathcal{E}_y}{i_x B_z}, \qquad (1.8)$$

is by far the best-known of the thermogalvanomagnetic coefficients. Isothermal conditions are required for measuring R_H.

A transverse electric field also results if there is a temperature gradient dT/dx in the presence of a magnetic field B_z. This is the Nernst effect. The Nernst coefficient,

$$|N| = \frac{\mathcal{E}_y}{B_z} \Big/ \frac{dT}{dx}, \qquad (1.9)$$

is the transverse equivalent of the Seebeck coefficient and should be measured in the absence of current flow.

The transverse equivalent of the Peltier effect is the Ettingshausen effect, which appears when one imposes a current and allows a transverse temperature gradient. The associated heat flow is a plausible means of electronic refrigeration, particularly at low temperatures. The Ettingshausen coefficient is defined by

$$|P| = \frac{1}{i_x B_z} \frac{dT}{dy}. \qquad (1.10)$$

1.2 Thermoelectric and Thermomagnetic Phenomena

As transverse equivalents of α and Π, N and P obey a relationship, similar to (1.7). This is the Bridgman relationship,

$$P\lambda = NT.\tag{1.11}$$

The Righi–Leduc effect is the appearance of a transverse temperature gradient due to an imposed longitudinal gradient. The corresponding coefficient,

$$|S| = \frac{1}{B_z}\frac{dT}{dy} \bigg/ \frac{dT}{dx},\tag{1.12}$$

should be measured in open circuit conditions.

Table 1.1 summarizes the definitions of the thermoelectric and thermomagnetic transport coefficients and the experimental conditions required for measuring them. Isothermal conditions, such as required for measuring the Peltier coefficient, can be difficult to obtain. Equations (1.7) and (1.11) are means around these difficulties.

Table 1.1. Transport coefficients in isotropic conductors

Name of coefficient	Symbol	Definition[a]	Conditions
Electrical resistivity ρ		$\dfrac{\mathscr{E}}{i_x}$	$i_y = i_z = 0, \nabla T = 0$
Thermal conductivity	λ	$-w_x \bigg/ \dfrac{dT}{dx}$	$i = 0, \dfrac{dT}{dy} = \dfrac{dT}{dz} = 0$
Seebeck	α	$\mathscr{E} \bigg/ \dfrac{dT}{dx}$	$i = 0, \dfrac{dT}{dy} = \dfrac{dT}{dz} = 0$
Peltier	Π	$\dfrac{w_x}{i_x}$	$i_y = i_z = 0, \nabla T = 0$
Hall	R_H	$\dfrac{\mathscr{E}}{i_x B_z}$	$i_y = i_z = 0, \nabla T = 0$
Nernst	N	$\dfrac{\mathscr{E}}{B_z} \bigg/ \dfrac{dT}{dx}$	$i = 0, \dfrac{dT}{dy} = \dfrac{dT}{dz} = 0$
Ettingshausen	P	$\dfrac{1}{i_x B_z}\dfrac{dT}{dy}$	$i_y = i_z = 0, \dfrac{dT}{dx} = \dfrac{dT}{dz} = 0$
Righi–Leduc	S	$\dfrac{1}{B_z}\dfrac{dT}{dy} \bigg/ \dfrac{dT}{dx}$	$i = 0, \dfrac{dT}{dz} = 0$

[a] i is the electric current per unit area of cross section; w is the rate of heat flow per unit area; T is the temperature; $\mathscr{E}$ is the electric field; B is the magnetic field; when B is nonzero, it lies in the z direction

1.3 Peltier Cooling and the Thermoelectric Figure of Merit

For refrigeration, a thermocouple consists of a "p" branch with a positive α and an "n" branch with a negative α (Fig. 1.5). These two branches are joined by a metal interconnect with low α, which we take as zero. The two legs of each couple and the many couples in a thermoelectric device are connected thermally in parallel and electrically in series. Such an arrangement allows one to achieve adequate heat pumping and convenient total resistance. We lose no generality in analyzing a single couple. And with regard to key results, there is also no loss of generality in assuming that the branches have constant cross-sections along their lengths.

For our analysis, we assume that the only electrical resistance is that of the thermocouple branches. We also assume that there is zero thermal resistance between the ends of the branches and the heat source and sink and that the only paths for transferring heat between the source and sink are the thermocouple branches, that is, we ignore conduction via the ambient, convection, and radiation. Finally, we assume that α, ρ, and λ do not vary with temperature.

Our objective is to find the coefficient of performance as a function of the temperature difference between the source and the sink. The coefficient of performance is defined as the ratio of the rate at which heat is extracted from the source to the rate of expenditure of electrical energy.

The origin of the Peltier effect resides in the transport of heat by an electric current. This heat flow changes abruptly at a junction between two dissimilar materials, liberating or absorbing heat. Within each branch, though, the Peltier current conducts heat, rather than generating heat. The total heat flow (positive heat flow is from the source to the sink) within each leg is

$$Q_\text{p} = \alpha_\text{p} IT - \lambda_\text{p} A_\text{p} \frac{dT}{dx}, \qquad Q_\text{n} = -\alpha_\text{n} IT - \lambda_\text{n} A_\text{n} \frac{dT}{dx}, \tag{1.13}$$

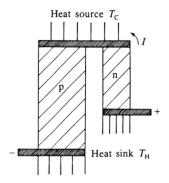

Fig. 1.5. Single-couple refrigerator

1.3 Peltier Cooling and the Thermoelectric Figure of Merit

where A is the cross-sectional area and (1.7) was used to insert α. The Peltier heat flow is positive in each branch because α_n is negative. Heat is removed from the source at the rate

$$Q_C = (Q_p + Q_n)|_{x=0}. \tag{1.14}$$

Within the branches, the rate of generation of heat per unit length from the Joule effect is $I^2\rho/A$. To balance this heat supply, there is a non constant thermal gradient:

$$-\lambda_p A_p \frac{d^2T}{dx^2} = \frac{I^2 \rho_p}{A_p}, \qquad -\lambda_n A_n \frac{d^2T}{dx^2} = \frac{I^2 \rho_n}{A_n}. \tag{1.15}$$

Because we have assumed that the Thomson coefficient is zero, the thermoelectric current does not supply or extract heat within the branches.

The boundary conditions are $T = T_C$ at $x = 0$, and $T = T_H$ at $x = L_{p,n}$, so that (1.15) gives

$$\lambda_{p,n} A_{p,n} \frac{dT}{dx} = -\frac{I^2 \rho_{p,n}\left(x - \frac{1}{2}L_{p,n}\right)}{A_{p,n}} + \frac{\lambda_{p,n} A_{p,n} (T_H - T_C)}{L_{p,n}}, \tag{1.16}$$

where $\Delta T = T_H - T_C$. Substituting in (1.13) and then using (1.14) leads to the net heat pumping rate,

$$Q_C = (\alpha_p - \alpha_n) I T_C - K \Delta T - \frac{1}{2} I^2 R, \tag{1.17}$$

where the total thermal conductance (parallel arrangement) and electrical resistance (series arrangement) are

$$K = \frac{\lambda_p A_p}{L_p} + \frac{\lambda_n A_n}{L_n}, \qquad R = \frac{L_p \rho_p}{A_p} + \frac{L_n \rho_n}{A_n}. \tag{1.18}$$

Despite the asymmetrical temperature gradient, the diffusive heat flow is only the total thermal conductance multiplied by the average gradient, and exactly one-half of the total Joule heating is seen as an extra load for the thermoelectric current.

The electrical power consumed by the thermocouple is not simply the Joule power. The external current source must also work against the Seebeck voltage. The total power consumption is

$$W = I\left[(\alpha_p - \alpha_n)\Delta T + IR\right]. \tag{1.19}$$

The two voltage terms in (1.19) can be separated experimentally because the Seebeck term dissipates slowly after the current is switched off. (We note that the separation of these two voltages is the basis of the Harman method for determining the thermoelectric coefficients of a sample, as will be discussed in detail in Chap. 4. The coefficient of performance ϕ is

$$\phi = \frac{Q_C}{W} = \frac{(\alpha_p - \alpha_n)IT_C - K\Delta T - \frac{1}{2}I^2 R}{I\left[(\alpha_p - \alpha_n)\Delta T + IR\right]}. \tag{1.20}$$

If there were no irreversible effects, we would find $\phi = T_C/\Delta T$, the Carnot limit. For a given ΔT, ϕ depends on the current, and two special values of the current correspond to maximum heat pumping and maximum efficiency. The current that yields the maximum heat pumping also produces the maximum ΔT. The figure of merit, $Z = (\alpha_p-\alpha_n)^2/RK$, the mean temperature T_M, and $\gamma = (1+ZT_M)^{1/2}$ for the couple are important parameters of a thermoelectric device, that are summarized in Table 1.2.

The results in Table 1.2 for ΔT, I, and ϕ do not change when a number of couples is linked together into a device, whereas Q is proportional to the number of couples. In practice, the A/L ratios of the couple branches are chosen to place the optimal currents in a desirable range, and the number of couples is matched to the projected thermal load. To extend the range of ΔT's available, it is necessary to construct multistage devices (Section 5.5.4).

T_H and T_C are all one needs to calculate ΔT_{max} and ϕ_{max} as functions of ZT_M, the dimensionless figure of merit for a couple. Figures 1.6 and 1.7 show some of these curves for different values of ZT_M and T_H/T_C ratios. ZT_M is approximately 0.8 for commercial heat sink materials at room temperature and operating near ΔT_{max}. From either figure, it is clear that reaching from room temperature to $T < 200$ K with single-stage devices will require significantly better materials.

Table 1.2. Ideal performance benchmarks of a thermoelectric couple

	Q_{max}	ϕ_{max}	ΔT_{max}
ΔT			$Z(T_C)^2/2$
I	$(\alpha_p-\alpha_n)T_C/R$	$(\alpha_p-\alpha_n)\Delta T/[R(\gamma-1)]$	$(\alpha_p-\alpha_n)T_C/R$
V	$(\alpha_p-\alpha_n)T_H$	$(\alpha_p-\alpha_n)\Delta T\gamma/(\gamma-1)$	$(\alpha_p-\alpha_n)T_H$
Q_C	$K\Delta T[Z(T_C)^2/2\Delta T-1]$	$K\Delta T[ZT_C-\Delta T(\gamma+1)/2T_M-1]$	0

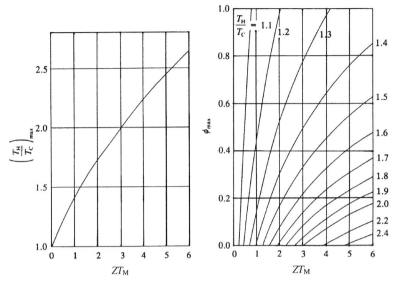

Fig. 1.6. Maximum ratio of hot to cold junction temperature as a function of the dimensionless figure of merit

Fig. 1.7. Maximum coefficient of performance as a function of the dimensionless figure of merit for different ratios of hot to cold junction temperature

Z for a couple is not a fixed quantity, but depends on the relative dimensions of the branches. It is maximized when the product RK is minimized, which occurs when

$$\frac{L_n A_p}{L_p A_n} = \left(\frac{\rho_p \lambda_n}{\rho_n \lambda_p}\right)^{\frac{1}{2}}. \tag{1.21}$$

If (1.21) is satisfied, the figure of merit becomes

$$Z = \frac{(\alpha_p - \alpha_n)^2}{\left[(\lambda_p \rho_p)^{1/2} + (\lambda_n \rho_n)^{1/2}\right]}. \tag{1.22}$$

In selecting thermocouple materials, one aims to maximize Z as defined by (1.22). In general, this can lead to selecting materials different from chose simply selected by choosing the best p-type and n-type materials according to their individual z's:

$$z_{p,n} = \frac{\alpha_{p,n}^2}{\rho_{p,n} \lambda_{p,n}}. \tag{1.23}$$

At most temperatures of interest, the thermoelectric properties of the best available p-type and n-type materials are similar. In this case, Z for a couple is approximately the average of z_p and z_n, and it is practical to optimize the materials separately. However, at low temperatures, this is not the situation. It is possible to tailor the p-type material for operation around 200 K to a greater extent than possible for the n-type material. Here, the materials become somewhat dissimilar.

At even lower temperatures, it becomes advantageous to switch to Bi–Sb alloys for the n-type material, making the properties of the two branches grossly different. As the liquid nitrogen temperature range is approached, the best thermocouples are composed of n-type Bi–Sb alloys and a passive (α, $\rho = 0$) superconducting leg. Equation (1.22) suggests that $Z = Z_n$ for such a couple, but this is incorrect because one cannot set $A_p = 0$, as (1.21) requires. The superconducting branch cross-section must be large enough to keep the current density below the critical value and the properties and dimensions of the Bi–Sb leg adjusted to maximize the Z of the couple.

1.4 Efficiency of Thermoelectric Power Generation

In a thermoelectric power generation device, diffusive heat flow and the Peltier effect are additive. Both reduce the imposed temperature gradient. The resistive voltage drop of the device also detracts from the voltage available from the Seebeck effect. With these differences in mind, the equivalent of (1.20) for power generation efficiency η, is given by

$$\eta = \frac{W}{Q_H} = \frac{I\left[(\alpha_p - \alpha_n)\Delta T - IR\right]}{K\Delta T + (\alpha_p - \alpha_n)IT_H - \frac{1}{2}I^2 R}, \tag{1.24}$$

where W is the power delivered to an external load and Q_H is positive for heat flow from the source to the sink. Again, there is a value of I that maximizes η for a fixed device design. In thermoelectric power generation applications, operating at η_{max} (Table 1.3) entails matching the device and load resistances. Recalling that the Carnot efficiency of a generator is $\Delta T/T_H$, it is easy to see from the expressions in Table 1.3 that the efficiency of thermoelectric generation for $ZT < 1$ is less than 30% of the Carnot limit.

Table 1.3. Ideal performance standards for thermoelectric generation

	W_{max}	η_{max}
I	$(\alpha_p - \alpha_n)\Delta T / 2R$	$(\alpha_p - \alpha_n)\Delta T / R(\gamma + 1)$
R_{load}	R	γR
W	$[(\alpha_p - \alpha_n)\Delta T]^2 / 4R$	$\gamma[(\alpha_p - \alpha_n)\Delta T]^2 / R(\gamma + 1)^2$
η	$Z\Delta T / (4 + ZT_H + ZT_M)$	$(\gamma - 1)\Delta T / [(\gamma + 1)T_H - \Delta T]$

1.5 Ettingshausen Cooling and the Thermomagnetic Figure of Merit

The basic device for electronic cooling by the Ettingshausen effect is shown in Fig. 1.8. There is fairly close correspondence between the expressions for cooling power and coefficient of performance obtained by using the Ettingshausen and Peltier effects [1.3]. However, the Ettingshausen refrigerator requires only one material, so the issue of matching n-type and p-type transport properties does not arise.

The transport coefficients involved in describing the basic Ettingshausen cooling device are the Ettingshausen (P) and Nernst (N) coefficients, along with λ and ρ. For P, N and λ the conditions given in Table 1.1 are valid for this discussion. However, to achieve the greatest degree of similarity with Peltier cooling, we use ρ as measured with zero transverse electric currents, zero transverse thermal gradients, and zero longitudinal heat flow, rather than a zero longitudinal thermal gradient [1.4, 1.5]. This is referred to as the adiabatic ρ, and its use leads to the adiabatic thermomagnetic figure of merit Z_E.

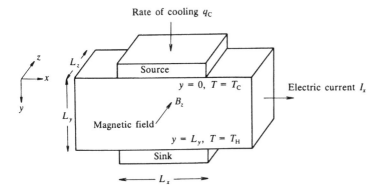

Fig. 1.8. Basic thermomagnetic refrigerator

The thermomagnetic element of cross-section $L_y L_z$ is in contact with a source and sink over a length L_x. A current, I_x, is passed in the x direction with a magnetic field B_z in the z direction. Heat is removed from the source at a rate Q_C and transferred to the sink along the y direction. We assume that N, ρ, and λ are independent of temperature.

The Ettingshausen heat flow per unit area is given by $P\lambda B_z I_x L_x/L_y$, which, from (1.11), can be written as $NB_z I_x TL_x/L_y$. Thus, the rate of heat flow in the y direction is given by

$$Q = \frac{NB_z I_x TL_x}{L_y} - \lambda L_x L_z \frac{dT}{dy}, \tag{1.25}$$

a form similar to (1.13). Again, Joule heating gives rise to a non constant temperature gradient. The Joule power per unit length in the y direction is $(I_x)^2 \rho L_x/(L_z L_y)^2$. This leads to the thermomagnetic equivalent of (1.15),

$$\frac{I_x^2 \rho L_x}{L_z L_y^2} = -\lambda L_x L_z \frac{d^2 T}{dy^2}. \tag{1.26}$$

The boundary conditions $T = T_C$ at $y = 0$ and $T = T_H$ at $y = L_y$ give an intermediate result similar to (1.16). The cooling power is

$$Q_C = \frac{NB_z I_x T_C L_x}{L_y} - \frac{\lambda L_x L_z \Delta T}{L_y} - \frac{I_x^2 \rho L_x}{2 L_z L_y}. \tag{1.27}$$

The form is the same as that of (1.17) for Peltier cooling, where $NB_z L_x/L_y$ is in place of $(\alpha_p - \alpha_n)$, $\lambda L_x L_z/L_y$ in place of K and $\rho L_x/(L_y L_z)$ in place of R. Therefore, we can use Table 1.2 to predict ideal efficiencies and $\Delta T_{\max}$, and Z is replaced by

$$Z_E = \frac{(NB_z)^2}{\rho \lambda}. \tag{1.28}$$

Despite the analytical congruence between Ettingshausen and Peltier cooling, only the latter is a viable commercial technology. With available materials, the magnetic fields required for efficient Ettingshausen cooling are too large, except at low temperatures. Exploitation of some of the advantages of thermomagnetic cooling (e.g., the simplicity of device design and construction and the possibility of achieving greater ΔT's by altering the element shape) awaits the discovery of a material that can be used with good efficiency in the field of an inexpensive permanent magnet.

2. Transport of Heat and Electricity in Solids

2.1 Crystalline Solids

Condensed matter may be found in many different forms. For example, liquids, amorphous solids, organic materials, and ionic conductors have been considered as possible thermoelectric materials. The materials presently employed in devices, however, are inorganic (semiconductor) crystalline solids. Therefore these materials will form the basis of our treatment in this chapter, where we review the basic properties of solids, that are relevant to thermoelectric science and technology.

We suppose, then, that a solid body consists of one or more crystals; the atoms in each of them are regularly arranged in the form of a lattice. The crystal lattice is the framework along which the carriers of electric charge and thermal energy move. The vibrations of the lattice provide one of the principal mechanisms of heat transfer. Then, an understanding of crystals is a basic requirement.

There is a wide variety of crystal structures that may be assumed by an element or compound. The particular structure that is observed is governed to some extent by the nature of the binding between the atoms.

The simplest bonds arise from coulombic attraction between ionized atoms that bear charges of opposite sign. Sodium chloride is a well-known ionic solid. In this compound, each of the sodium atoms loses one electron to become positively charged. The chlorine atoms become negatively charged as each takes up an electron. The arrangement in the sodium chloride structure is stable when each positive or negative ion is surrounded symmetrically by six ions of the opposite sign. The sodium chloride, or rock salt, structure is shared by similar alkali halides and is shown in Fig. 2.1. If all of the atoms were identical, we would have a simple cubic lattice. Taking account of the two different types of atom, we see that the lattice is of the face-centered cubic type. Note that a repulsive force that becomes very strong as the spacing becomes smaller opposes the coulombic attraction between oppositely charged ions. At equilibrium atomic spacing, the attractive and repulsive forces balance.

It is useful to examine the periodic table, shown as Table 2.1, and discuss the different possible bonding arrangements between elements. Group IV contains the elements silicon and germanium, which, like one form of carbon and gray tin, have a diamond structure. In this structure, each atom shares the four electrons in its

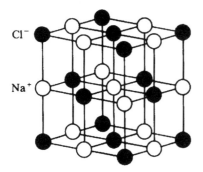

Fig. 2.1. The rock salt structure

outer shell with its four nearest neighbors. Thus, the binding between each pair of atoms involves two electrons and is termed covalent. Unlike an ionic bond, a covalent bond is strongly directional so that, in the diamond structure, the four nearest neighbors of each atom lie at the corners of a tetrahedron. (In Part II of the text, we will show how this well-known bonding scheme allows the formation of some very interesting compounds.) In covalently bonded crystals, the electrons are concentrated near the lines that join neighboring atoms, whereas in ionic crystals, the electrons form almost spherical clouds around the negative ions. Figure 2.2 shows the atomic arrangement in the diamond structure and its face-centered cubic lattice.

Table 2.1. Periodic table of the elements. The symbols for the elements are accompanied by their atomic numbers and atomic masses

Group I	Group II				Transition elements							Group III	Group IV	Group V	Group VI	Group VII	Group 0
$_1$H 1.008																	$_2$He 4.003
$_3$Li 6.941	$_4$Be 9.012											$_5$B 10.81	$_6$C 12.01	$_7$N 14.01	$_8$O 16.00	$_9$F 19.00	$_{10}$Ne 20.18
$_{11}$Na 22.99	$_{12}$Mg 24.31											$_{13}$Al 26.98	$_{14}$Si 28.09	$_{15}$P 30.97	$_{16}$S 32.45	$_{17}$Cl 35.45	$_{18}$A 39.95
$_{19}$K 39.10	$_{20}$Ca 40.08	$_{21}$Sc 44.96	$_{22}$Ti 47.90	$_{23}$V 50.94	$_{24}$Cr 52.00	$_{25}$Mn 54.94	$_{26}$Fe 55.85	$_{27}$Co 58.93	$_{28}$Ni 58.71	$_{29}$Cu 63.55	$_{30}$Zn 65.37	$_{31}$Ga 69.72	$_{32}$Ge 72.92	$_{33}$As 74.92	$_{34}$Se 78.96	$_{35}$Br 79.90	$_{36}$Kr 83.80
$_{37}$Rb 85.47	$_{38}$Sr 87.62	$_{39}$Y 88.91	$_{40}$Zr 91.22	$_{41}$Nb 92.91	$_{42}$Mo 95.94	$_{43}$Tc 98.91	$_{44}$Ru 101.1	$_{45}$Rh 102.9	$_{46}$Pd 106.4	$_{47}$Ag 107.9	$_{48}$Cd 112.4	$_{49}$In 114.8	$_{50}$Sn 118.7	$_{51}$Sb 121.8	$_{52}$Te 127.6	$_{53}$I 126.9	$_{54}$Xe 131.3
$_{55}$Cs 132.9	$_{56}$Ba 137.3	57-71	$_{72}$Hf 178.5	$_{73}$Ta 180.9	$_{74}$W 183.9	$_{75}$Re 186.2	$_{76}$Os 190.2	$_{77}$Ir 192.2	$_{78}$Pt 195.1	$_{79}$Au 197.0	$_{80}$Hg 200.6	$_{81}$Tl 204.4	$_{82}$Pb 207.2	$_{83}$Bi 209.0	$_{84}$Po (210)	$_{85}$At (210)	$_{86}$Rn (222)
$_{87}$Fr (223)	$_{88}$Ra 226.0	89-103															
Lanthanides			$_{57}$La 138.9	$_{58}$Ce 140.1	$_{59}$Pr 140.9	$_{60}$Nd 144.2	$_{61}$Pm (147)	$_{62}$Sm 150.4	$_{63}$Eu 152.0	$_{64}$Gd 157.3	$_{65}$Tb 158.9	$_{66}$Dy 162.5	$_{67}$Ho 164.9	$_{68}$Er 167.3	$_{69}$Tm 168.9	$_{70}$Yb 173.0	$_{71}$Lu 175.0
Actinides			$_{89}$Ac (227)	$_{90}$Th 232.0	$_{91}$Pa 231.0	$_{92}$U 238.0	$_{93}$Np 237.0	$_{94}$Pu (242)	$_{95}$Am (243)	$_{96}$Cm (248)	$_{97}$Bk (247)	$_{98}$Cf (251)	$_{99}$Es (254)	$_{100}$Fm (253)	$_{101}$Md (256)	$_{102}$No (254)	$_{103}$Lr (257)

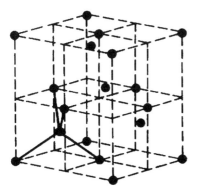

Fig. 2.2. The diamond structure. The face-centered cubic lattice is indicated by the broken lines; two atoms are associated with each lattice point. The bonds between one atom and its four neighbors are indicated by solid lines

The atoms to the left of Group IV are termed electropositive because they tend to give up, or donate, electrons, thereby becoming positive ions. Likewise, the atoms to the right of Group IV are termed electronegative and tend to take, or accept, electrons. The alkali halides exemplify the principle that compounds between strongly electropositive and strongly electronegative elements contain ionic bonds, whereas when the electronegativity difference is small, covalent bonding predominates. The nature of the bonding has a considerable bearing on the transport properties of a solid.

In a metallic bond, as in a covalent bond, atoms are bound together by shared electrons. However, in this case, the participating electrons are less influenced by s–p hybridization, and the electron density is more uniform. The number of nearest neighbors is no longer determined by considerations of valence, and many metals form structures in which the atoms are as near to one another as possible. These are face-centered cubic and close-packed hexagonal structures.

The fourth type of bond is much weaker than the three that we have already considered. This is the van der Waals bond between neutral atoms that arises from their small fluctuating dipole moments. This bond is particularly important in condensed inert gases (group 0 in the periodic table) because the other types of bonding are then absent. It is also of some significance in certain thermoelectric materials, notably those that have layered structures, as will be discussed in detail in Chap. 5.

It has been customary for solid-state scientists to direct their attention primarily to elements and compounds that form cubic crystal structures. In such substances, the thermoelectric properties are isotropic, that is, they are independent of crystal orientation. However, thus far the best materials for thermoelectric refrigeration form noncubic crystals, so that the thermoelectric properties are anisotropic. The particular crystal structures and the nature of the anisotropy will be discussed later. Note that

2.2 Heat Conduction by the Lattice

2.2.1 Pure Crystals

The fact that metals are good conductors of heat can be attributed to the transport of heat by the charge carriers. Of course, heat can also be conducted through electrical insulators. The thermal conductivity of diamond, for example, exceeds that of any metal [2.2]. It is, in fact, the transfer of vibrational energy from one atom to the next that constitutes this process.

The atoms in a crystal do not vibrate independently. Rather, the vibrations are such that elastic waves continually pass backward and forward through the crystal. An increase in temperature should lead to an increase in the amplitude of a given wave, as well as redistribution of the relative amplitudes for different wavelengths. This concept was used by Debye [2.3] to determine the specific heat of a solid. Although Debye's theory is now known to have shortcomings, it is still regarded as a useful first step toward the complete description of the thermal properties of solids, and we will use it here.

Debye represented the crystal as an elastic continuum. When boundary conditions are imposed, only certain modes of vibration are possible. For example, if the surfaces of the crystal are to correspond to nodes, this condition can be realized only for waves in a particular direction if they have particular wavelengths. In a true continuum, the total number of modes of vibration would be infinite. Debye showed that the discrete nature of atoms sets a lower limit to the wavelength. Therefore, one should take account of only the 3 $\mathcal{N}$ modes of lowest frequency, where $\mathcal{N}$ is the number of atoms per unit volume. This restriction on the number of vibrational modes gives a result at high temperatures that is consistent with the Dulong and Petit law, which states that the specific heat per atom is equal to $3k_B$, where k_B is Boltzmann's constant. The number of modes, dw, per unit volume that have frequencies between υ and $\upsilon + d\upsilon$ is given by

$$dw = \frac{2\pi \upsilon^2 d\upsilon}{v^3}, \qquad (2.1)$$

where v is an appropriate average value for the speed of sound. In a simple continuum, the speed of sound is independent of frequency, but, even in an isotropic solid, it will be different for longitudinal and transverse waves.

2.2 Heat Conduction by the Lattice

We must really distinguish between the phase velocity, defined by $2B\lambda/q_L$, and the group velocity, defined by $2Bd\lambda/dq_L$. Here, the wave number q_L, is equal to $2B/\gamma$, where γ is the wavelength. The plot of frequency against wave number, or the dispersion curve, is linear through the origin for a continuum but becomes nonlinear at high frequencies in real materials. If more than one atom is associated with each lattice point, the dispersion curve also has separate branches, that have one or more optical branches to each acoustic branch. Figure 2.3 shows a schematic dispersion curve for a diatomic linear lattice. In three dimensions, there will be three acoustic branches (one for the longitudinal vibrations and one for each polarization of the transverse vibrations) and $3(n-1)$ optical branches, where n is the number of atoms per primitive cell. The group velocity, determined by the slope of the dispersion curve, is of particular interest because it is related to the rate of energy transport. It clearly depends on frequency or wave number and also differs considerably for acoustic and optical modes.

The limit of $3\mathcal{N}$ on the number of modes means that there is an upper frequency limit υ_D, which may be obtained by integrating (2.1). υ_D is given by the relationship

$$\frac{4\pi \upsilon_D^3}{v^3} = 3\mathcal{N}. \tag{2.2}$$

Debye's success in explaining the general behavior of specific heat as a function of temperature was due to his use of quantum theory. He showed that the average energy W in a mode of frequency υ is

$$W = h\upsilon \left[\exp\left(\frac{h\upsilon}{k_B T}\right) - 1 \right]^{-1}, \tag{2.3}$$

where h is Planck's constant. With the frequency limit set by (2.2), Eqs. (2.1) and (2.3) allow one to determine the internal energy of the crystal. Then, by differentiating the internal energy with respect to temperature, the specific heat at constant volume is

$$c_V = 9\mathcal{N}k_B \left(\frac{T}{\theta_D}\right)^3 f_D\left(\frac{\theta_D}{T}\right), \tag{2.4}$$

where the quantity θ_D, known as the Debye temperature, is defined by

$$\theta_D = \frac{h\upsilon_D}{k_B} \tag{2.5}$$

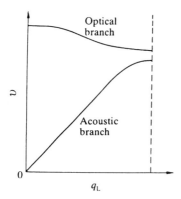

Fig. 2.3. Schematic dispersion curve for a diatomic linear lattice

and

$$f_D\left(\frac{\theta_D}{T}\right) = \int_0^{\theta_D/T} \frac{x^4 \exp x}{(\exp x - 1)^2} dx. \qquad (2.6)$$

Debye's expression (2.4) for specific heat indicates that $c_V/\mathcal{N}k_B$ should be a universal function of the reduced temperature T/θ_D. Thus the plot of c_V versus T/θ_D (Fig. 2.4) should apply for all solids according to the continuum model, where θ_D is a different constant for each material. In fact, experimental values of specific heat closely follow the behavior illustrated in Fig. 2.4. Debye's theory has been so successful that it is usual to fit (2.4) to the experimental data by regarding θ_D as a temperature-dependent parameter.

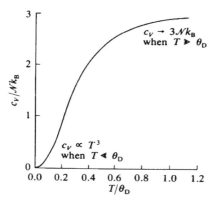

Fig. 2.4. Dependence of specific heat at constant volume on termperature, according to Debye's theory

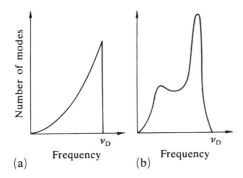

Fig. 2.5. Vibrational spectra for **(a)** a Debye continuum and **(b)** a simple cubic lattice

The relative accuracy of (2.4) does not indicate that Debye's theory gives a particularly good description of the vibrations in a real crystal. Therefore, specific heat studies are not a good way to determine the distribution of vibrational modes. The vibrational spectra can, however, be found by the neutron scattering method of Brockhouse [2.4] and others. As an illustration, in Fig. 2.5, we compare the vibrational spectrum for a continuum with that of a simple cubic lattice. Although we shall use the concepts of the Debye theory, the nature of the approximations should always be borne in mind.

The first requirement of any theory of lattice heat transport is that it should explain the observation of Eucken [2.5] that, for pure dielectric crystals, thermal conductivity varies inversely with absolute temperature. Eucken's rule has been confirmed for many materials, provided that the temperature is not much less than θ_D.

Debye [2.6] attempted to use his continuum model to explain Eucken's $1/T$ law but found that the thermal conductivity is infinite. It is only by taking account of the anharmonic nature of the forces in real crystals that thermal conductivity takes on finite values. If an atom is displaced from its equilibrium position by an infinitesimal amount, the restoring force is proportional to the displacement, i.e., it is harmonic. As the displacement becomes larger, the expression for the restoring force includes higher order terms. Then, the force is partially anharmonic. In effect, the elastic constants of the substance vary from point to point according to the local displacements of the atoms. Variations in either the density or the elastic properties are responsible for scattering the vibrational waves and, thus, for limiting the value of thermal conductivity.

Peierls [2.7] showed how the anharmonicity of the lattice vibrations could be taken into account. He introduced the concept of phonon wave packets that arise from the quantization of vibrational waves. We shall refer to the wave packets simply as phonons, and they may be regarded as the energy carriers that are responsible for heat conduction by the lattice. A limit on thermal conductivity is set by phonon collisions that do not conserve momentum. We may draw on the kinetic theory of gases to express the lattice conductivity λ_L in terms of the mean free path l_t of the phonons. Thus,

$$\lambda_L = \frac{1}{3} c_V v l_t, \qquad (2.7)$$

where the specific heat c_V is defined for unit volume and, as before, v is the speed of sound and l_t is the mean free path length.

In pure crystals at high temperatures, phonons are scattered predominantly by other phonons. The most important processes involve three phonons and can be of two kinds. In the normal or N-processes both energy and momentum are conserved. In Umklapp or U-processes, momentum (or more strictly wave number) is not conserved. The N-processes lead to a redistribution of phonons but no thermal resistance to first order. Therefore, the U-processes directly control high-temperature thermal conductivity.

The difference between the N- and U-processes is indicated in Fig. 2.6. Two-dimensional representations of wave-vector space are shown; the squares correspond to all values of the phonon wave vector set by the frequency limit v_D. The maximum x or y component of the wave vector is equal to half the side of the square. The region of wave-vector space that contains the allowed values for the wave vector is called the Brillouin zone.

In the N-process, two phonons of wave vector q_1 and q_2, interact to form a third phonon of wave vector q_3 such that

$$\boldsymbol{q_1} + \boldsymbol{q_2} = \boldsymbol{q_3}. \qquad (2.8)$$

On the other hand, in the U-process, the sum of the wave vectors q_1 and q_2 would produce a resultant outside the Brillouin zone, if (2.8) were applied. Thus,

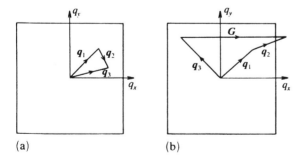

Fig. 2.6. Two-dimensional representation of **(a)** N-process and **(b)** U-process

one subtracts the vector $\mathbf{G}$, the reciprocal lattice vector, to produce a resultant wave vector $\mathbf{q}_3$ that lies within the allowed zone. The rule for the conservation of a wave vector for the U-process is

$$\mathbf{q}_1 + \mathbf{q}_2 = \mathbf{q}_3 + \mathbf{G}. \tag{2.9}$$

Peierls showed that the increasing probability of U-processes with a rise in temperature explains Eucken's $1/T$ law. However, at low temperatures, it becomes exceedingly difficult for any U-processes to occur. When $T \ll \theta_D$, there are virtually no phonons where wave vectors are comparable to the half-width of the Brillouin zone and very few with even half that value. The only real possibility for finding a U-process is when two phonons have wave vectors marginally greater than a quarter of the Brillouin zone width, provided that they happen to have more or less the same direction. The number of such phonons is proportional to $\exp(-\theta_D/2T)$, so one might expect that l_t at low temperatures is proportional to $[\exp(-\theta_D/aT)]^{-1}$, where $a \approx 2$. The exponential variation of l_t should have a major influence on the dependence of thermal conductivity on temperature.

Although there is no doubt that the inverse temperature dependence of thermal conductivity, when $T > \theta_D$, will always change to an exponential dependence for $T \ll \theta_D$ in a sufficiently pure and perfect large crystal, in practice, it is often difficult to observe such behavior. It is usually found that scattering by all kinds of internal defects and by the external crystal boundaries masks phonon–phonon scattering at low temperatures. In fact, one can sometimes take advantage of these other phonon-scattering processes to improve thermoelectric materials. Therefore, it is important for us to consider lattice conductivity in less than perfect crystals.

2.2.2 Scattering of Phonons by Defects

As the temperature of a dielectric crystal is lowered, its thermal conductivity rises to a maximum value and then falls so as to approach zero at 0 K. At the lowest temperatures, thermal conductivity is proportional to T^3, which is the same variation with temperature as that of specific heat. This implies that l_t must have reached a limiting value. Casimir [2.8] showed that this value is comparable to the dimensions of the crystal and can be attributed to scattering by phonons at the crystal boundaries. In a polycrystalline sample, we expect that the maximum free path length of the phonons is set by the grain size. It is even found that amorphous solids behave as if heat conduction arises from phonons whose free path lengths comparable to the interatomic spacing.

One might expect boundary scattering to have an effect on the lattice conductivity of crystalline materials only when the temperature is low enough for l_t of phonons to be very large. However, as will be shown later, it is sometimes possible to observe boundary scattering at ordinary temperatures [2.9].

Now, we discuss scattering by the various point defects that are found in real crystals. Examples of such defects are the local variations of elasticity and density that occur in solid solutions between isomorphous crystals (e.g., in silicon–germanium alloys) or in an element or compound that contains foreign impurities or vacancies. It is even possible to observe scattering of phonons due to the density variations associated with the different isotopes in naturally occurring elements [2.10].

The treatment of the scattering of phonons by point defects presents us with a difficulty. The relaxation time τ_I for this form of scattering should be proportional to $1/v^4$. This means that point defect scattering does not have a significant effect on low-frequency phonons. The Thermal conductivity at low temperatures then becomes exceedingly large because U-processes also have little effect in this region. That this phenomenon does not occur can be explained if one takes account of the redistribution of phonons by N-processes into modes for which the scattering is stronger.

Callaway [2.11] developed a satisfactory method of including N-processes. In Callaway's treatment, one uses the relaxation time τ, rather than l_t; the two quantities are related through $l_t = v\tau$. The basic principle is that non-momentum-conserving processes cause the perturbed phonons to relax toward equilibrium distribution, whereas N-processes cause them to relax toward a nonequilibrium distribution. If all of the scattering processes were non-momentum-conserving, the overall relaxation time could be obtained simply as the reciprocal of the sum of the reciprocal relaxation times of the individual processes. Callaway successfully obtained a multiplying factor that corrects for the momentum-conserving behavior of N-processes.

Suppose that there are two different relaxation times τ_R for the processes that change the momentum and τ_N for the N-processes. The rate at which the phonon-distribution function N changes through the various scattering effects is

$$\left(\frac{dN}{dt}\right)_{scatter} = \frac{N_0 - N}{\tau_R} + \frac{N_N - N}{\tau_N}, \qquad (2.10)$$

where N_0 is the equilibrium distribution function and N_N is the distribution function to which N-processes on their own would lead. At the same time, the temperature gradient ∇T changes the distribution function according to

$$\left(\frac{dN}{dt}\right)_{diffusion} = -\mathbf{v} \cdot \nabla T \frac{\partial N}{\partial T}, \qquad (2.11)$$

where v is the sound velocity in the direction of the phonon wave vector q_L. Because the diffusion and scattering processes must balance,

$$\frac{N_0 - N}{\tau_R} + \frac{N_N - N}{\tau_N} - v \cdot \nabla T \frac{\partial N}{\partial T} = 0. \tag{2.12}$$

Now, the N-processes must lead to a distribution that carries momentum against the temperature gradient. A suitable distribution is

$$N_N = \left[\exp\left(\frac{\hbar\omega - q_L \cdot l}{k_B T}\right) - 1 \right]^{-1}, \tag{2.13}$$

instead of the usual Bose–Einstein distribution function

$$N_0 = \left[\exp\left(\frac{\hbar\omega}{k_B T}\right) - 1 \right]^{-1}. \tag{2.14}$$

It is convenient to use the angular frequency ω, which is equal to $2\pi v$, and $\hbar = h/2\pi$. In (2.13), l is a constant vector in the direction of ∇T where $q_L \cdot l \ll \hbar\omega$. In effect, (2.13) differs from (2.14) in that the frequency ω is changed by $q_L \cdot l / \hbar$. Then, we find, that

$$N_N - N_0 = \frac{q_L \cdot l}{k_B T} \frac{\exp x}{(\exp x - 1)^2}, \tag{2.15}$$

where $x \equiv \hbar\omega / k_B T$. Also

$$\frac{\partial N}{\partial T} = \frac{\hbar\omega}{k_B T^2} \frac{\exp x}{(\exp x - 1)^2}, \tag{2.16}$$

so that

$$N_N - N_0 = \frac{q_L \cdot l \, T}{\hbar\omega} \frac{\partial N}{\partial T}. \tag{2.17}$$

Substituting (2.17) in (2.12) gives

2. Transport of Heat and Electricity in Solids

$$(N_0 - N)\left(\frac{1}{\tau_R} + \frac{1}{\tau_N}\right) - \left(\mathbf{v} \cdot \nabla T - \frac{\mathbf{q}_L \cdot l T}{\hbar \omega \tau_N}\right)\frac{\partial N}{\partial T} = 0. \tag{2.18}$$

Because l should be proportional to the temperature gradient, we may write

$$l = -\frac{\hbar}{T}\beta v^2 \nabla T, \tag{2.19}$$

where β is a constant whose dimensions are time. Also, using the Debye model to express v as ω/q_L, (2.18) becomes

$$(N_0 - N)\left(\frac{1}{\tau_R} + \frac{1}{\tau_N}\right)\left(1 + \frac{\beta}{\tau_N}\right)^{-1} - \mathbf{v} \cdot \nabla T \frac{\partial N}{\partial T} = 0. \tag{2.20}$$

This means that the behavior is exactly the same as it would be if there were an effective relaxation time τ_{eff} given by

$$\frac{1}{\tau_{\text{eff}}} = \left(\frac{1}{\tau_R} + \frac{1}{\tau_N}\right)\left(1 + \frac{\beta}{\tau_N}\right)^{-1} = \frac{1}{\tau_c}\left(1 + \frac{\beta}{\tau_N}\right)^{-1}. \tag{2.21}$$

Here, τ_c is the relaxation time that we would expect if the N-processes did not conserve wave vector because then we would merely add the reciprocal relaxation times for all of the scattering processes. We see that τ_c must be multiplied by $(1 + \beta/\tau_N)$ to yield τ_{eff}.

The constant β is found by using the fact that N-processes conserve the wave vector. In other words,

$$\int \frac{N_N - N}{\tau_N} \mathbf{q}_L \mathrm{d}^3 q_L = \int \frac{4\pi \omega^2}{v^3} \frac{N_N - N}{\tau_N} q_L \mathrm{d}\omega = 0. \tag{2.22}$$

Because $N_N - N = (N_N - N_0) + (N_0 - N)$ is proportional to $(\beta - \tau_{\text{eff}})\, \partial N / \partial T$, (2.22) can be written as

$$\int_0^{\theta_D/T}\left(\frac{\beta}{\tau_N} - \frac{\tau_c}{\tau_N} - \frac{\beta \tau_c}{\tau_N^2}\right)\omega^4 \frac{\exp x}{(\exp x - 1)^2}\mathrm{d}x = 0. \tag{2.23}$$

Thus, we find that

$$\beta = \int_0^{\theta_D/T} \frac{\tau_c}{\tau_N} x^4 \frac{\exp x}{(\exp x - 1)^2} dx \bigg/ \int_0^{\theta_D/T} \frac{1}{\tau_N}\left(1 - \frac{\tau_c}{\tau_N}\right) x^4 \frac{\exp x}{(\exp x - 1)^2} dx. \quad (2.24)$$

The general determination of β is difficult, but Callaway showed that a simplification of the expression is possible in certain cases. For example, if the scattering by the imperfections is very strong and the relaxation time $\tau_I \ll \tau_N$, it is valid to use the approximation $1/\tau_{\text{eff}} \approx 1/\tau_I + 1/\tau_N$. Another approximation may be made at low temperatures for fairly pure crystals, when $\tau_I \gg \tau_N$ and U-processes can be neglected. However, when we are dealing with thermoelectric materials, the high temperature approximation used by Parrott [2.12] is of interest. The relaxation times for both U- and N-processes are expected to be proportional to ω^2, whereas for point defects, τ_I is proportional to ω^{-4}. Suppose, then, that the relaxation times are expressed as $1/\tau_I = A\omega^4$, $1/\tau_U = B\omega^2$, $1/\tau_N = C\omega^2$, where $A, B,$ and C are constants for a given sample. Also, when $T \gg \theta_D$, $x \ll 1$ for the whole phonon spectrum, so that $x^2 \exp x/(\exp x - 1)^2 \approx 1$. Then it is possible to obtain analytical solutions for the integrals involved in β and to express λ_L for the crystal with point imperfections in terms of the value λ_0 that would be found for a perfectly pure crystal. The expression is

$$\frac{\lambda_L}{\lambda_0} = \left(1 + \frac{5k_0}{9}\right)^{-1}\left[\frac{\tan^{-1} y}{y} + \left(1 - \frac{\tan^{-1} y}{y}\right)^2 \left(\frac{y^4(1+k_0)}{5k_0} - \frac{y^2}{3} - \frac{\tan^{-1} y}{y}\right)^{-1}\right], (2.25)$$

where k_0, equal to C/B, is a measure of the relative strengths of the N- and U-processes. The quantity y is defined by

$$y^2 = \left(\frac{\omega_D}{\omega_0}\right)^2 \left(1 + \frac{5k_0}{9}\right)^{-1}, \quad (2.26)$$

where ω_0 is given by

$$\left(\frac{\omega_0}{\omega_D}\right)^2 = \frac{k_B}{2\pi^2 v \lambda_0 \omega_D A}. \quad (2.27)$$

The value of k_0 may be determined experimentally by measuring the thermal conductivity of one pure crystal and one that contains imperfections. Thereafter, the same value of k_0 can be used for crystals with other concentrations of defects.

Note that the approximations that hold strictly for $T \gg \theta_D$ are often used with reasonable success for the less stringent condition $T \geq \theta_D$.

2.3 Band Theory of Solids

2.3.1 Energy Bands in Metals, Insulators, and Semiconductors

In the classical free electron theory of metals, it was supposed that the atomic nuclei and all except the outer electrons are in fixed positions, whereas the outer electrons form a "gas" that can be described by the familiar kinetic theory. In particular, a mean thermal energy of $3k_BT/2$ was ascribed to each of the free electrons; the energy is distributed according to the Maxwell–Boltzmann law.

The classical free electron theory was consistent with Ohm's law and predicted the right order of magnitude for the electrical resistivity. It also explained the observed relationship between the electrical and thermal conductivity in metals (the Wiedemann–Franz law), but it failed badly in certain other respects. A most obvious shortcoming was its prediction of a substantial specific heat from the electrons, whereas it is observed experimentally that electronic specific heat is negligible at ordinary temperatures. This and other failures of the classical theory can be eliminated, however, if we take account of the fact that the free-electron gas should obey Fermi–Dirac statistics rather than Maxwell–Boltzmann statistics. That is to say, classical theory must be replaced by quantum theory [2.13]. The following rules must be followed:

(1) Electrons can have spin quantum numbers of $+1/2$ or $-1/2$ but are otherwise indistinguishable from one another.
(2) Electrons have discrete values for their energy and momentum.
(3) No more than one electron of a given spin can reside in one of these discrete states.

Sommerfeld showed that the number of electron states permitted per unit volume within the energy range from E to $E + dE$ is given by

$$g(E)dE = \frac{4\pi(2m)^{3/2} E^{1/2} dE}{h^3}, \tag{2.28}$$

where m is the mass of a free electron. The probability that a particular state will be filled is given by the Fermi distribution function

$$f_0(E) = \left[\exp\left(\frac{E - \zeta}{k_B T}\right) + 1\right]^{-1}, \tag{2.29}$$

where ζ is called the Fermi energy, which has a value such that

$$\int_0^\infty f_0(E)g(E)dE = n, \tag{2.30}$$

where n is the total number of free electrons per unit volume. At a temperature of absolute zero, Fermi energy has the value

$$\zeta_0 = \frac{h^2}{2m}\left(\frac{3n}{8\pi}\right)^{2/3}. \tag{2.31}$$

All of the states with $E < \zeta_0$ are filled with electrons, whereas all of the states with $E > \zeta_0$ are empty. This, of course, is in contrast to the result of classical theory that all of the electrons have zero energy at 0 K. At ordinary temperatures, $\zeta \gg k_B T$ for all metals, which means that $\zeta \approx \zeta_0$.

Figure 2.7(a) shows a schematic plot of the Fermi distribution function against energy. It is only when the energy lies within a few $k_B T$ of the Fermi energy that the occupation probability differs significantly from either one or zero. Typically, ζ is of the order of 100 times $k_B T$ at ordinary temperatures for metals, but it is certainly not true that $\zeta \gg k_B T$ for semimetals or semiconductors. The distribution of the electrons in a metal is shown as a function of energy in Fig. 2.7(b). Provided that T remains much smaller than ζ/k_B, the distribution is little changed from that at 0 K, when the temperature is raised. This explains why the electronic contribution to specific heat is so small.

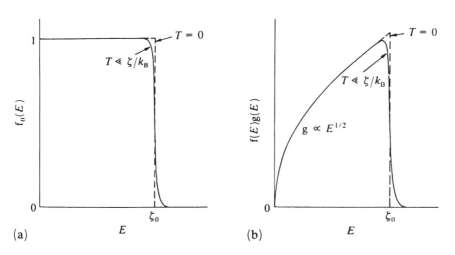

Fig. 2.7. Schematic plots of **(a)** Fermi distribution function against energy for a typical metal and **(b)** distribution of electrons with energy in a metal

In (2.31) we see that the value of the Fermi energy at the temperature of absolute zero can become small if the free-electron concentration n is much less than normally found in a metal. Then it is possible to obtain the condition $k_B T \gg \zeta_0$ at ordinary temperatures, whence (2.29) reduces to

$$f_0(E) = \exp\left(-\frac{E-\zeta}{k_B T}\right). \tag{2.32}$$

This is, in fact, the Maxwell–Boltzmann distribution function that is used in the classical kinetic theory of gases. When n is small enough for (2.32) to be applicable, the electron gas is said to be nondegenerate. On the other hand, when n is so large that $k_B T \ll \zeta_0$, we have what is called a degenerate electron gas. It turns out that thermoelectric materials tend to lie between these two extremes.

In an electronic transport process, the electrons move from state to state in steps of no more than about $k_B T$. This movement requires an empty state within about $k_B T$ of the original filled state. Suitable pairs of states are found only near the Fermi energy. In other words, in a metal, very few of the electrons can actually contribute to the transport of charge or energy. The value of the observed electrical resistivity then requires a much longer free path between collisions than was expected from the classical theory. The free electron theory of metals does not explain why the electron free path length should be so great, nor does it account for the observed variation of the resistivity with temperature. Another curious fact is that the sign of the charge carriers, as revealed by the sign of the thermoelectric effects or the Hall coefficient, is sometimes positive, whereas free electrons always bear a negative charge. These observations can be explained if we introduce a further concept, the band theory of solids.

Electrons in a crystalline solid do not move in free space but in a regular array of atoms. In the so-called one-electron model, the motion of each electron is considered separately from that of the others. The effect of the remaining electrons and the atomic nuclei is to provide a potential that varies periodically in space. Bloch showed that the wave equation for an electron, which is situated in such a potential, has only certain solutions. Characteristically, these solutions fall within certain bands of energy that can be separated from one another by forbidden bands.

The most useful way of presenting the distribution of the allowed states in terms of energy and wave vector is the band-structure diagram. Such a diagram portrays the different permissible energies for each value of the wave vector. Note that all of the information can be contained in a diagram that includes wave vectors within certain prescribed limits. The wave vectors lie within the region of reciprocal space, already referred to for phonons, known as the Brillouin zone or, more strictly, the first Brillouin zone. Wave vectors that lie outside the first Brillouin zone can be brought back into it by adding a reciprocal lattice vector without any change in their physical significance. When the band-structure diagram is

2.3 Band Theory of Solids

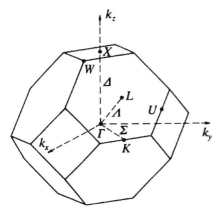

Fig. 2.8. First Brillouin zone for the face-centered cubic lattice

represented for wave vectors that lie in the first Brillouin zone, it is said to use the reduced zone scheme.

We may illustrate these ideas most easily for a crystal with cubic symmetry, so reference will be made to the semiconductor silicon. Figure 2.8 shows the appropriate first Brillouin zone, that for a face-centered cubic lattice, where certain points of symmetry are indicated by conventional labels. The band-structure diagram for silicon with wave vectors along the ΓX and ΓL directions is shown schematically in Fig. 2.9. To avoid complication, only the allowed energy bands that are associated with the ordinary semiconducting properties are shown, but it must be appreciated that there are other bands at higher and lower energies.

In Fig. 2.9, the bands labeled VB(H) and VB(L) overlap one another but are separated from the band labeled CB by a forbidden gap. The maxima of VB(H) and VB(L) occur at the center Γ of the Brillouin zone, and the minimum of CB occurs at a point in reciprocal space on the line joining Γ and X. Because there are six equivalent X points, symmetry demands that there must be the same number of equivalent minima. Thus, silicon is said to be a multivalley semiconductor. The forbidden gap E_g between the bands VB and CB has a width of 1.1 eV, and is said to be indirect because the band extrema are at different wave-vector values. There is also a larger direct energy gap.

In a free-electron gas, the energy E varies as the square of the wave vector k. This clearly cannot be generally true for energy bands such as those shown in Fig. 2.9, although, close to the maxima and minima, the bands are parabolic. This means that, if we measure the energy and the wave vector from the band edge, $E \propto k^2$ for small values of E and k. Thus we may write

$$E = \frac{h^2 k^2}{8\pi^2 m^*}, \tag{2.33}$$

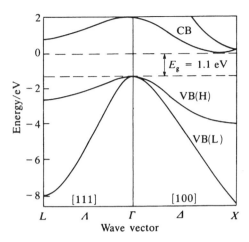

Fig. 2.9. Schematic energy band-structure diagram for silicon

which is the same expression as for a free-electron gas, except that the electronic mass, m, has been replaced by m^*, the effective mass. The same effective mass may be inserted in (2.28) to yield a density of states near the band maximum or minimum of

$$g(E)dE = \frac{4\pi(2m^*)^{3/2}|E|^{1/2}\,d|E|}{h^3}. \tag{2.34}$$

In these cases, the interaction between the electrons and the crystal lattice is taken into account by using m^*.

The effective mass may be defined quite generally by the expression

$$\frac{1}{m^*} = \frac{1}{\hbar^2}\frac{d^2 E}{dk^2}, \tag{2.35}$$

but it is only near the band edges that (2.33) and (2.34) can be usefully employed. It is of special interest to consider what happens when d^2E/dk^2 becomes negative. The crystal then behaves as if it contained negative carriers of negative mass, but it is convenient to regard these as positive carriers of positive mass, known as positive holes (usually just "holes"). This feature of band theory explains why the values of the thermoelectric and Hall coefficients are sometimes positive. When dealing with holes, E must be measured downward from the band edge. The Fermi energy ζ must also be measured downward from the same band edge and the function f(E) then represents the probability that a state is unoccupied by an electron, that is, occupied by a hole.

Energy-band theory has been particularly successful in accounting for the differences among metals, semiconductors, and insulators. In a metal, the Fermi energy lies within one or more of the allowed bands and well away from the band edge. The density of electron states is large in this region, and the electrical conductivity is high for such a partly filled band. On the other hand, in an electrical insulator, the Fermi level lies within the forbidden gap. The bands below the gap are completely full, so there are no free states into which electrons can move. The bands above are completely empty, so they cannot contribute to the conduction process either.

In fact, this last statement is not quite true, except at the temperature of absolute zero. At any temperature above 0 K, some electrons will be excited from the highest full band to the lowest empty band, so that there will be a finite, though very small, electrical conductivity. If the energy gap E_g is small enough, substantial numbers of electrons in the upper band and the same number of holes in the lower band will be excited in this way. The substance is then said to be an intrinsic semiconductor.

It is possible to turn semiconductor materials, which are essentially insulators in their pure states, into electrical conductors, by adding impurities. Sometimes the impurities give up electrons to the upper band and are said to be donors. Alternatively, the impurities may take up electrons from the lower band, so as to leave behind positive holes; these impurities are called acceptors. In both cases, the states associated with the impurities usually lie very close to the band edges, and excitation is virtually complete at ordinary temperatures. The materials in which the charge carriers originate from the addition of impurities are known as extrinsic semiconductors.

Figure 2.10 gives energy diagrams for the different types of nonmetals. Energy diagrams rather than band-structure diagrams are useful when we have no particular interest in the value of the wave vector. Note that the upper and lower allowed bands are called the conduction and valence band, respectively.

The essential difference between an insulator and an intrinsic semiconductor is just the width of the energy gap in terms of $k_B T$. Thus, an insulator at ordinary temperatures may become an intrinsic semiconductor at an elevated temperature. Likewise, as the temperature of an extrinsic semiconductor is raised, the charge carriers excited across the energy gap will eventually predominate over the carriers associated with the impurities. When this happens, the extrinsic semiconductor becomes intrinsic.

There are certain materials in which the valence and conduction bands just overlap one another. They are called semimetals, and they do not differ substantially in their properties from narrow-gap semiconductors. Adding donor or acceptor impurities can, for example, change their carrier concentration. This contrasts with the situation in ordinary metals in which changes of the electrical resistivity on adding impurities result from a reduction in the electronic free path length rather than from any alteration in the number of charge carriers.

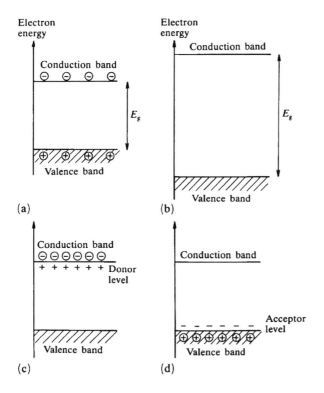

Fig. 2.10. Energy diagrams for nonmetals: (**a**) intrinsic semiconductor, (**b**) insulator, (**c**) extrinsic n-type semiconductor, (**d**) extrinsic p-type semiconductor

2.3.2 Electron-Scattering Processes

The transport properties of a conductor depend not on the concentration of charge carriers and also on the free path length between collisions. This would have an infinite value if the electrons or holes were moving in a perfectly periodic potential, because then they would never be scattered. However, in any real crystal, the electronic free path length becomes limited by scattering due to local distortions of the potential electronic function.

Transport coefficients will be derived from Boltzmann's equation, and the scattering processes will be described in terms of a relaxation time τ_e. If the electron distribution function is disturbed from its equilibrium value f_0, to f, where $|f-f_0| \ll f_0$, the system relaxes according to the equation

$$\left(\frac{df(E)}{dt}\right)_{scatter} = -\frac{f(E) - f_0(E)}{\tau_e}. \tag{2.36}$$

Hopefully, one can express the relaxation time in the form

$$\vartheta_e = \vartheta_0 E^r ,\qquad(2.37)$$

where τ_0 and r are constants. There may well be several scattering processes, with relaxation times $\tau_{e,1}$, $\tau_{e,2}$ ..., that should all be taken into account, whereupon

$$\frac{1}{\tau_e} = \frac{1}{\tau_{e,1}} + \frac{1}{\tau_{e,2}} + ...,\qquad(2.38)$$

but we shall usually suppose that one process predominates.

In a pure and perfect crystal, the carriers are scattered by the thermal vibrations of the crystal lattice. Both the acoustic and the optical modes can upset the periodicity of the potential. The acoustic modes perturb the lattice potential in two ways. The local changes of lattice parameter directly influence the energy-band structure, leading to so-called deformation potentials. Alternatively, if the atoms in the crystal are partly ionized, piezoelectric potentials result from their displacement. It is usually assumed that the deformation-potential scattering is much stronger than the piezoelectric scattering.

Similarly, optical vibrations can cause scattering in more or less the same two ways. In this case, it is usual to refer to the mechanisms as nonpolar and polar. The polar process that creates dipole moments which vary in time and space, may completely outweigh nonpolar processes.

When electrons and holes with a given energy are located at different points in the Brillouin zone, as in a multivalley semiconductor, lattice scattering may involve large changes of wave vector (intervalley scattering), as well as small changes (intravalley scattering). The two processes have different effects at low temperatures, but at high temperatures, when there are substantial numbers of phonons with large wave vectors, intervalley scattering becomes similar to intravalley scattering.

A number of other scattering processes become possible in imperfect and impure crystals. Impurity atoms act as scattering centers, particularly if they are ionized. Crystal defects, such as dislocations, can also produce a significant effect. Often, mixed crystals or solid solutions are used in thermoelectric materials, in which case alloy scattering can become important. Finally, remember that the charge carriers can scatter one another, though it is doubtful that carrier–carrier scattering is significant at the concentrations in which we are interested.

There are several excellent treatments of scattering processes ranging from an elementary account by Ioffe [2.14] to a very thorough treatment by Nag [2.15] for compound semiconductors. The discussion by Beer [2.16] is also relevant. In the next chapter, we shall consider the processes that are considered most important in thermoelectric materials.

2.4 Electron Transport in a Zero Magnetic Field

2.4.1 Single Bands

Now, we are in a position to derive general expressions for thermogalvanomagnetic coefficients. For now, though, we will restrict ourselves to the case of zero magnetic field. It will be supposed that we are dealing with one type of carrier, that resides in a single band of the parabolic form described by (2.34). It is also assumed that, although the charge carriers interact with the phonons, they have no significant effect on their distribution. In other words, phonons are regarded merely as scattering centers. Later, we shall consider the phonon-drag effect, in which this assumption is relaxed.

The Boltzmann equation for the steady state represents a balance between the effects of the fields and the scattering processes. Thus,

$$\frac{f(E) - f_0(E)}{\tau_e} = -\dot{\bm{k}} \cdot \frac{\partial f(E)}{\partial \bm{k}} - \dot{\bm{r}} \cdot \frac{\partial f(E)}{\partial \bm{r}}, \qquad (2.39)$$

where f and f_0 are the perturbed and unperturbed distribution functions, respectively, and $\bm{k}$ and $\bm{r}$ are the wave vector and position vector of the charge carriers. The dot above these vectors represents the partial derivative with respect to time. We must solve this equation to obtain $f - f_0$ in terms of the electric field and the temperature gradient. It will be assumed that the conductor is isotropic and that the fields and flows are all in the direction of the x axis.

If $|f - f_0| \ll f_0$, f can be replaced by f_0 on the right-hand side of (2.39). Then, bearing in mind that the Fermi distribution function will be dependent on displacement through both the electric field and the temperature gradient, we may show that

$$\frac{f(E) - f_0(E)}{\tau_e} = u \frac{\partial f_0(E)}{\partial E} \left(\frac{\partial \zeta}{\partial x} + \frac{E - \zeta}{T} \frac{\partial T}{\partial x} \right), \qquad (2.40)$$

where u is the velocity of the carriers in the x direction. This equation applies equally well for holes and electrons, provided that E and ζ are measured from the appropriate band edge and in the appropriate direction. Thus, electron energies are measured upward from the bottom of the conduction band, and hole energies downward from the top of the valence band.

The number of charge carriers per unit volume in the range of energy from E to $E + dE$ is $f(E)g(E)dE$; $g(E)$ is given by (2.34). Because the carriers, with charge $\mp e$, move in the x direction with a velocity u, the electric current density is

2.4 Electron Transport in Zero Magnetic Field

$$i = \mp \int_0^\infty e u f(E) g(E) dE. \qquad (2.41)$$

Here and in subsequent equations, the upper sign refers to electrons, and the lower sign to holes. The rate of flow of heat per unit cross-sectional area is

$$w = \int_0^\infty u(E - \zeta) f(E) g(E) dE, \qquad (2.42)$$

because $E - \zeta$ represents the total energy transported by a carrier. Note that the upper limit of the integrals in (2.41) and (2.42) suggests that the bands are infinitely wide; in practice, however, the function f(E) always falls to zero before E becomes very large.

Because there can be no flow of charge or heat when $f = f_0$, one can replace f by $(f - f_0)$ in the two previous equations. Also, in any case of interest to us, the thermal velocity of the charge carriers will always be much greater than any drift velocity. Thence, we may set

$$u^2 = \frac{2E}{3m^*}, \qquad (2.43)$$

because u^2 will be one-third of the mean square velocity that corresponds to the energy E. By inserting (2.43) in (2.41) and (2.42), we find that

$$i = \mp \frac{2e}{3m^*} \int_0^\infty g(E) \tau_e E \frac{\partial f_0(E)}{\partial E} \left(\frac{\partial \zeta}{\partial x} + \frac{E - \zeta}{T} \frac{\partial T}{\partial x} \right) dE, \qquad (2.44)$$

and

$$w = \pm \frac{\zeta}{e} i + \frac{2}{3m^*} \int_0^\infty g(E) \tau_e E^2 \frac{\partial f_0(E)}{\partial E} \left(\frac{\partial \zeta}{\partial x} + \frac{E - \zeta}{T} \frac{\partial T}{\partial x} \right) dE. \qquad (2.45)$$

To find an expression for the electrical conductivity σ, we set the temperature gradient $\partial T / \partial x$ equal to zero. The electric field is given by $\pm(\partial \zeta / \partial x) e^{-1}$, so that

$$\sigma = \frac{i}{\mathcal{E}} = -\frac{2e^2}{3m^*} \int_0^\infty g(E) \tau_e E \frac{\partial f_0(E)}{\partial E} dE. \qquad (2.46)$$

Alternatively, if the electric current is zero, (2.44) shows that

$$\frac{\partial \zeta}{\partial x}\int_0^\infty g(E)\tau_e E\frac{\partial f_0(E)}{\partial E}dE + \frac{1}{T}\frac{\partial T}{\partial x}\int_0^\infty g(E)\tau_e E(E-\zeta)\frac{\partial f_0(E)}{\partial E}dE = 0. \quad (2.47)$$

The specified condition is that for the definition of both the Seebeck coefficient and thermal conductivity. The Seebeck coefficient is equal to $(\partial \zeta / \partial x)\,[e\,(\partial T / \partial x)]^{-1}$ and is given by

$$\alpha = \pm\frac{1}{eT}\left[\zeta - \int_0^\infty g(E)\tau_e E^2 \frac{\partial f_0(E)}{\partial E}dE \bigg/ \int_0^\infty g(E)\tau_e E\frac{\partial f_0(E)}{\partial E}dE\right]. \quad (2.48)$$

The Seebeck coefficient is negative if the carriers are electrons and positive if they are holes.

The electronic thermal conductivity is given by $-w(\partial T/\partial x)^{-1}$ and is found from (2.45) and (2.47), so that

$$\lambda_e = \frac{2}{3m^*T}\left\langle \left[\int_0^\infty g(E)\tau_e E^2 \frac{\partial f_0(E)}{\partial E}dE\right]^2 \bigg/ \int_0^\infty g(E)\tau_e E\frac{\partial f_0(E)}{\partial E}dE \right.$$
$$\left. -\int_0^\infty g(E)\tau_e E^3 \frac{\partial f_0(E)}{\partial E}dE\right\rangle. \quad (2.49)$$

The Peltier coefficient Π can be found either from the Seebeck coefficient, using the appropriate Kelvin relationship, or as the ratio w/i, when $\partial T / \partial x = 0$.

All of the integrals that appear in (2.46), (2.48), and (2.49) have the same general form. They may be expressed conveniently as

$$K_s = -\frac{2T}{3m^*}\int_0^\infty g(E)\tau_e E^{s+1}\frac{\partial f_0(E)}{\partial E}dE$$
$$= -\frac{8\pi}{3}\left(\frac{2}{h^2}\right)^{3/2}(m^*)^{1/2}T\tau_0\int_0^\infty E^{s+r+3/2}\frac{\partial f_0(E)}{\partial E}dE, \quad (2.50)$$

where g and τ_e have been eliminated in terms of m^*, r, and τ_0 using (2.34) and (2.37). After integrating by parts,

$$\int_0^\infty E^{s+r+3/2}\frac{\partial f_0(E)}{\partial E}dE = -\left(s+r+\frac{3}{2}\right)\int_0^\infty E^{s+r+1/2}f_0(E)dE. \quad (2.51)$$

Thus, we may write

$$K_s = \frac{8\pi}{3}\left(\frac{2}{h^2}\right)^{3/2}(m^*)^{1/2}T\tau_0\left(s+r+\frac{3}{2}\right)(k_B T)^{s+r+3/2}F_{s+r+1/2}, \qquad (2.52)$$

where

$$F_n(\xi) = \int_0^\infty \xi^n f_0(\xi)\,d\xi, \qquad (2.53)$$

and the reduced energy ξ is equal to $E/k_B T$. Numerical values of the functions F_n, which are known as Fermi–Dirac integrals, are given in [2.17–2.19] for various values of η, the reduced Fermi energy, defined as $\zeta/k_B T$.

The expressions for the transport coefficients in terms of the integrals K_s are

$$\sigma = \frac{e^2}{T}K_0, \qquad (2.54)$$

$$\alpha = \pm\frac{1}{eT}\left(\zeta - \frac{K_1}{K_0}\right), \qquad (2.55)$$

and

$$\lambda_e = \frac{1}{T^2}\left(K_2 - \frac{K_1^2}{K_0}\right). \qquad (2.56)$$

These equations allow expressing the figure of merit z in terms of ζ, m^*, the relaxation time parameters, τ_0 and r, and λ_L. The lattice conductivity is involved because the total thermal conductivity λ is given by

$$\lambda = \lambda_L + \lambda_e. \qquad (2.57)$$

2.4.2 Degenerate and Nondegenerate Conductors

In general, it is necessary to use numerical methods to determine the transport properties from (2.54) to (2.57). However, if the Fermi energy is either much greater than or much less than zero, it is possible to use simple approximations to the Fermi distribution function.

First, we shall consider the nondegenerate approximation that is applicable when $\zeta/k_B T \ll 0$, that is, the Fermi level lies within the forbidden gap well away from the appropriate band edge. The approximation is very good when $\zeta < -4k_B T$ and is acceptable for many purposes when $\zeta < -2k_B T$. It is necessary that the Fermi level should be much further from the opposite band edge, if minority carriers are to be neglected. From a practical point of view, the independent variable is the carrier concentration, because this can usually be controlled directly in a semiconducting material. Here, it will be more convenient to regard the Fermi energy as the variable quantity.

When $\zeta/k_BT \ll 0$, the Fermi–Dirac integrals become

$$F_n(\eta) = \exp(\eta)\int_0^\infty \xi^n \exp(-\xi)d\xi = \exp(\eta)\Gamma(n+1), \tag{2.58}$$

where the gamma function has the property

$$\Gamma(n+1) = n\Gamma(n). \tag{2.59}$$

When n is an integer, $\Gamma(n + 1) = n!$, and the gamma function can be calculated for half-integral values of n from the fact that $\Gamma(1/2) = \pi^{1/2}$.

In terms of the gamma function, the transport integrals take the form

$$K_s = \frac{8\pi}{3}\left(\frac{2}{h^2}\right)^{3/2} (m^*)^{1/2} T\tau_0 (k_BT)^{s+r+3/2}\Gamma\left(s+r+\frac{5}{2}\right)\exp(\eta). \tag{2.60}$$

Thus, the electrical conductivity of a nondegenerate semiconductor is

$$\sigma = \frac{8\pi}{3}\left(\frac{2}{h^2}\right)^{3/2} e^2(m^*)^{1/2} \tau_0 (k_BT)^{r+3/2}\Gamma\left(r+\frac{5}{2}\right)\exp(\eta). \tag{2.61}$$

It is convenient to express electrical conductivity as

$$\sigma = ne\mu, \tag{2.62}$$

where n is the carrier concentration and μ is the mobility, which is the average drift speed of the carriers in a unit electric field. The concentration of the carriers is given by

$$n = \int_0^\infty f(E)g(E)dE = 2\left(\frac{2\pi m^* k_BT}{h^2}\right)^{3/2} \exp(\eta), \tag{2.63}$$

so that the mobility is

$$\mu = \frac{4}{3\pi^{1/2}} \Gamma\left(r + \frac{5}{2}\right) \frac{e\tau_0 (k_B T)^r}{m^*}. \tag{2.64}$$

One of the consequences of the nondegenerate approximation is that the mobility in (2.64) does not directly depend on the Fermi energy. It must be remembered that τ_0 and r, may well depend on the carrier concentration, unless lattice scattering predominates.

It is seen from (2.63) that the carrier concentration has the value that one would expect if there were just $2(2\pi m^* k_B T/h)^{3/2}$ states located at the band edge. Thus, this quantity is known as the effective density of states. It has different values for different bands because it depends on m^*.

The Seebeck coefficient of a nondegenerate semiconductor from (2.55) and (2.60) is

$$\alpha = \pm \frac{k_B}{e}\left[\eta - \left(r + \frac{5}{2}\right)\right]. \tag{2.65}$$

One may regard $-\eta$ as the reduced potential energy of the carriers and $(r + 5/2)$ as the reduced kinetic energy that is transported by the current. Thus, the Peltier coefficient, equal to αT, represents the total energy transport per unit charge.

It is conventional to describe λ_e in terms of the Lorenz number L, defined as $\lambda_e/\sigma T$. Then, from (2.54) and (2.56),

$$L = \frac{1}{e^2 T^2}\left(\frac{K_2}{K_0} - \frac{K_1^2}{K_0^2}\right). \tag{2.66}$$

When the nondegenerate approximation is applicable

$$L = \left(\frac{k}{e}\right)^2 \left(r + \frac{5}{2}\right). \tag{2.67}$$

We see that L is independent of the Fermi energy, provided that the exponent r in the energy dependence of the relaxation time is constant.

Now we turn to the degenerate condition when $\zeta/k_B T \gg 0$. This means that the Fermi level is well above the conduction-band edge for electrons or well below the valence-band edge for holes. In other words, the conductor is metallic. The Fermi–Dirac integrals are expressed in the form of a rapidly converging series

$$F_n(\eta) = \frac{\eta^{n+1}}{n+1} + n\eta^{n-1}\frac{\pi^2}{6} + n(n-1)(n-2)\eta^{n-3}\frac{7\pi^4}{360} + \ldots, \tag{2.68}$$

In the degenerate approximation, one uses only as many of the terms in the series as needed to yield a finite or nonzero value for the appropriate parameter.

The electrical conductivity of a degenerate conductor is found by employing only the first term in the series,

$$\sigma = \frac{8\pi}{3}\left(\frac{2}{h^2}\right)^{3/2} e^2 (m^*)^{1/2} \tau_0 \zeta^{r+3/2}. \tag{2.69}$$

On the other hand, if only the first term in (2.68) were employed the Seebeck coefficient would be zero, which is consistent with the fact that most metals have exceedingly small values of α. To obtain a nonzero value for the Seebeck coefficient, the first two terms are used. Thus,

$$\alpha = \mp \frac{\pi^2}{3}\frac{k_B}{e}\frac{\left(r+\frac{3}{2}\right)}{\eta}. \tag{2.70}$$

The first two terms are also needed to obtain the Lorenz number, which is given by

$$L = \frac{\pi^2}{3}\left(\frac{k_B}{e}\right)^2. \tag{2.71}$$

This shows that the Lorenz number should be the same for all metals and, in particular, it should not depend on the scattering law for the charge carriers. These features agree fully with the well-established Wiedemann–Franz–Lorenz law which states that all metals have the same ratio of thermal to electrical conductivity and that this ratio is proportional to the absolute temperature. The only significant exceptions are for some pure metals at very low temperatures and for certain alloys in which the small value of λ_e means that the lattice contribution cannot be neglected.

2.4.3 Multiband Effects

So far it has been assumed that the carriers in only one band contribute to transport processes. Now we consider what happens when there is more than one type of carrier. The most important examples are the cases of mixed and intrinsic semiconductors in which there are significant contributions from both the electrons in the conduction band and holes in the valence band. Another example is found when there are comparable numbers of carriers of the same sign with different

2.4 Electron Transport in Zero Magnetic Field

effective masses, such as the light and heavy holes in silicon. The problem will be discussed for two types of carriers (represented by the subscripts 1 and 2).

There will be contributions to the electric current density i from both types of carriers. These contributions may be expressed in terms of partial electrical conductivities and partial Seebeck coefficients. Thus,

$$i_1 = \sigma_1\left(\mathcal{E} - \alpha_1 \frac{\partial T}{\partial x}\right), \qquad i_2 = \sigma_2\left(\mathcal{E} - \alpha_2 \frac{\partial T}{\partial x}\right). \tag{2.72}$$

The partial coefficients may be found by using the single-band approach that has already been outlined. Note that, when the electric field and the temperature gradient are in the same direction, the thermoelectric emf opposes $\mathcal{E}$ for a positive Seebeck coefficient and assists $\mathcal{E}$ when the Seebeck coefficient is negative. If the temperature gradient is zero,

$$i = i_1 + i_2 = (\sigma_1 + \sigma_2)\,\mathcal{E}, \tag{2.73}$$

so that the electrical conductivity is simply

$$\sigma = \sigma_1 + \sigma_2. \tag{2.74}$$

If we set the electric current equal to zero, which requires that i_1 and i_2 be equal and opposite,

$$(\sigma_1 + \sigma_2)\mathcal{E} = (\alpha_1\sigma_1 + \alpha_2\sigma_2)\frac{\partial T}{\partial x}. \tag{2.75}$$

Thus, the Seebeck coefficient is

$$\alpha = \frac{\alpha_1\sigma_1 + \alpha_2\sigma_2}{\sigma_1 + \sigma_2}. \tag{2.76}$$

As might be expected, the result is an average of the two partial Seebeck coefficients, weighted according to the partial electrical conductivities.

Now, if we examine thermal flow, the heat-flux densities due to the two carrier types are

$$w_1 = \alpha_1 T i_1 - \lambda_{e,1}\frac{\partial T}{\partial x}, \qquad w_2 = \alpha_2 T i_2 - \lambda_{e,2}\frac{\partial T}{\partial x}, \tag{2.77}$$

where the contributions from the Peltier effect have been expressed in terms of the partial Seebeck coefficients, using Kelvin's relationship. Because thermal conductivity is defined in the absence of an electric current, again i_1 and i_2 must be equal and opposite. We find that

$$i_1 = -i_2 = \frac{\sigma_1 \sigma_2}{\sigma_1 + \sigma_2}(\alpha_2 - \alpha_1)\frac{\partial T}{\partial x}. \tag{2.78}$$

By substitution in (2.77), total heat flux density is

$$w = w_1 + w_2 = -\left[\lambda_{e,1} + \lambda_{e,2} + \frac{\sigma_1 \sigma_2}{\sigma_1 + \sigma_2}(\alpha_2 - \alpha_1)^2 T\right]\frac{\partial T}{\partial x}, \tag{2.79}$$

and

$$\lambda_e = \lambda_{e,1} + \lambda_{e,2} + \frac{\sigma_1 \sigma_2}{\sigma_1 + \sigma_2}(\alpha_2 - \alpha_1)^2 T. \tag{2.80}$$

In (2.80), the remarkable point is that the total electronic thermal conductivity is not merely the sum of the partial conductivities $\lambda_{e,1}$ and $\lambda_{e,2}$. The presence of the third term results from the fact that Peltier heat flows can take place when there is more than one type of carrier, even when the total electric current is zero. The difference, $\alpha_2 - \alpha_1$, between the partial Seebeck coefficients is, of course, greatest when the two carriers are holes and electrons, respectively. Thus, we expect λ_e to be significantly greater than $\lambda_{e,1} + \lambda_{e,2}$ for an intrinsic semiconductor. The additional contribution, known as the bipolar thermodiffusion effect [2.20], is actually observed most easily in semiconductors that have a small energy gap. Then, although $\alpha_2 - \alpha_1$ is smaller than it would be for a wide-gap semiconductor, intrinsic conduction takes place with reasonably large values for the partial electrical conductivities σ_1 and σ_2. If σ_1 and σ_2 are too small, the electronic component of the thermal conductivity, even with bipolar thermodiffusion, is masked by the lattice component.

2.5 Effect of a Magnetic Field

2.5.1 One Type of Carrier

Now, we consider the application of a transverse magnetic field B to a conductor with charge carriers in a single band. The force on an electron, which is equal to $-e\mathscr{E}$ in a zero magnetic field, becomes $-(e\mathscr{E} + e\mathbf{u} \times \mathbf{B})$, where $\mathbf{u}$ is the electron velocity. Thus, the rate of change of the wave vector with time becomes

2.5 Effect of a Magnetic Field

$$k = -\frac{e}{\hbar}\left(\mathcal{E} + u \times B\right). \tag{2.81}$$

Then, it is assumed [2.21] that the perturbed distribution function f(E) may be written as

$$f(E) = f_0(E) - u \cdot c \frac{\partial f_0(E)}{\partial E}, \tag{2.82}$$

where c is a function of E. Thus, it can be shown that the Boltzmann equation (2.39) becomes

$$\frac{f(E)-f_0(E)}{\tau_e} = \frac{\partial f_0(E)}{\partial E} u \cdot \left[\nabla \zeta + \left(\frac{E-\zeta}{T}\right)\nabla T\right] - \frac{e}{\hbar}\frac{\partial f_0(E)}{\partial E} u \times B \frac{\partial u \cdot c}{\partial k}. \tag{2.83}$$

This equation, in turn, can be written as

$$\frac{f(E)-f_0(E)}{\tau_e} = \frac{\partial f_0(E)}{\partial E} u \cdot \left[\nabla \zeta + \left(\frac{E-\zeta}{T}\right)\nabla T - \frac{e}{m^*} B \times c\right]. \tag{2.84}$$

Thus, if (2.82) is used,

$$-\frac{c}{\tau_e} = \nabla \zeta + \left(\frac{E-\zeta}{T}\right)\nabla T - \frac{e}{m^*} B \times c. \tag{2.85}$$

It is assumed that the conductor is isotropic, that the magnetic field is in the z direction, and the electric field and the temperature gradient are in the x–y plane. We may therefore resolve c into its components with magnitude c_x and c_y in the x and y directions, respectively. Thus,

$$-\frac{c_x}{\tau_e} = \frac{\partial \zeta}{\partial x} + \frac{E-\zeta}{T}\frac{\partial T}{\partial x} + \frac{e}{m^*} B_z c_y,$$

$$-\frac{c_y}{\tau_e} = \frac{\partial \zeta}{\partial y} + \frac{E-\zeta}{T}\frac{\partial T}{\partial y} - \frac{e}{m^*} B_z c_x. \tag{2.86}$$

When these simultaneous equations are solved for c_x and c_y, we find that

$$c_x = \frac{k_1 - \beta_M k_2}{1 + \beta_M^2}, \qquad c_y = \frac{k_2 + \beta_M k_1}{1 + \beta_M^2}, \tag{2.87}$$

where

$$\beta_M = \frac{e\tau_e B_z}{m^*}, \tag{2.88}$$

$$k_1 = -\tau_e\left(\frac{\partial \zeta}{\partial x} + \frac{E-\zeta}{T}\frac{\partial T}{\partial x}\right), \quad k_2 = -\tau_e\left(\frac{\partial \zeta}{\partial y} + \frac{E-\zeta}{T}\frac{\partial T}{\partial y}\right). \tag{2.89}$$

Now, the electric current density and the heat flux density can be determined by using (2.41) and (2.42):

$$i_x = e\int_0^\infty u_x^2 c_x g(E)\frac{\partial f_0(E)}{\partial E}dE, \tag{2.90}$$

$$i_y = e\int_0^\infty u_y^2 c_y g(E)\frac{\partial f_0(E)}{\partial E}dE, \tag{2.91}$$

and

$$w_x = -\int_0^\infty u_x^2 c_x (E-\zeta)g(E)\frac{\partial f_0(E)}{\partial E}dE, \tag{2.92}$$

$$w_y = -\int_0^\infty u_y^2 c_y (E-\zeta)g(E)\frac{\partial f_0(E)}{\partial E}dE, \tag{2.93}$$

These equations may be compared with (2.44) and (2.45), which hold in the absence of a magnetic field. It will be seen that the integrals K_s are no longer applicable. Instead, we must use integrals of the form

$$\mathcal{K}_s = \frac{2T}{3m^*}\int_0^\infty \frac{g(E)\tau_e E^{s+1}}{1+\beta_M^2}\frac{\partial f_0(E)}{\partial E}dE, \tag{2.94}$$

and

$$\mathcal{H}_s = \frac{2T}{3m^*}\int_0^\infty \frac{\beta_M g(E)\tau_e E^{s+1}}{1+\beta_M^2}\frac{\partial f_0(E)}{\partial E}dE. \tag{2.95}$$

Note that at small magnetic fields, when $\beta_M \ll 1$, $\mathcal{K}_s$ tends toward K_s, and $\mathcal{H}_s$ tends toward

$$H_s = -\frac{2T}{3m^*}\int_0^\infty \beta_M g(E)\tau_e E^{s+1} \frac{\partial f_0(E)}{\partial E} dE. \tag{2.96}$$

The expressions for the various thermogalvanomagnetic coefficients may be found in the work of Tsidil'kovskii [2.22] and have been conveniently tabulated by Putley [2.21]. The coefficients of special interest to us are ρ, α, λ_e, and the Nernst coefficient, defined under appropriate conditions.

The isothermal electrical resistivity ρ^i is obtained by setting $\partial T/\partial x$, $\partial T/\partial y$, and i_y equal to zero so that

$$\rho^i = \frac{\mathcal{H}_0 T}{e^2\left(\mathcal{H}_0^2 + \mathcal{H}_0'^2\right)}. \tag{2.97}$$

The definition of the thermomagnetic figure of merit z_E involves the adiabatic electrical resistivity ρ^a. The two resistivities are related [2.23] through

$$\rho^a = \frac{\rho^i}{1+z_E T}. \tag{2.98}$$

The Seebeck coefficient in the magnetic field B_z is found by setting, i_x, i_y, and $\partial T/\partial y$ equal to zero.

$$\alpha = \mp\frac{1}{eT}\left(\frac{\mathcal{H}_0 \mathcal{H}_1 + \mathcal{H}_0' \mathcal{H}_1'}{\mathcal{H}_0^2 + \mathcal{H}_0'^2} - \zeta\right). \tag{2.99}$$

The same set of conditions is applicable for λ_e, which is given by

$$\lambda_e = \frac{1}{T^2}\left(\mathcal{H}_2 + \frac{\mathcal{H}_0 \mathcal{H}_1^2 - 2\mathcal{H}_0' \mathcal{H}_1 \mathcal{H}_1' - \mathcal{H}_0 \mathcal{H}_1'^2}{\mathcal{H}_0^2 + \mathcal{H}_0'^2}\right), \tag{2.100}$$

and also for the Nernst coefficient

$$N = \frac{1}{eTB_z}\left(\frac{\mathcal{H}_0 \mathcal{H}_1' - \mathcal{H}_0' \mathcal{H}_1}{\mathcal{H}_0^2 + \mathcal{H}_0'^2}\right). \tag{2.101}$$

Although seemingly, it is not required for determining the figure of merit, the Hall coefficient R_H is required in the course of basic studies on thermoelectric materials. In general, it may be expressed as

$$R_\mathrm{H} = \mp \frac{A}{ne}. \tag{2.102}$$

Like the Seebeck coefficient, the Hall coefficient is negative for conduction by electrons and positive for hole conduction. The numerical constant A is equal to unity for degenerate conductors and quite generally in a high magnetic field. However, it has a different value for a nondegenerate conductor when the relaxation time is energy-dependent. For example, if the scattering parameter r is equal to $-1/2$, A is equal to $3\pi/8$. For many purposes, it is a good approximation to replace A by unity.

2.5.2 Bipolar Thermomagnetic Effects

A consideration of the expressions for the transport coefficients in a zero magnetic field shows that one should avoid having electrons and holes present simultaneously, if a high thermoelectric figure of merit is sought. However, the bipolar effects are by no means undesirable when we are looking for a high thermomagnetic figure of merit. It is important, then, that we discuss the thermogalvanomagnetic properties of intrinsic semiconductors.

Let us first consider the origin of the thermomagnetic effects in extrinsic and intrinsic semiconductors. This will be done with reference to the Ettingshausen effect. It is supposed that the charge carriers in the extrinsic semiconductor, shown in Fig. 2.11(a), are electrons and that their scattering becomes stronger as their energy rises. Then, the drift speed of the less energetic carriers will exceed that of those that are more energetic. Consequently, the effect of a transverse magnetic field (directed outward from the diagram) is to exert a stronger downward force on the less energetic electrons. The electric field due to the Hall effect ensures that there is no overall transverse flow of charge, so that there is a balance between the "hotter" electrons moving upward and the "colder" electrons moving down. The heating and cooling effects at the top and bottom of the specimen, respectively, are not usually very strong and, of course, disappear altogether if the relaxation time of the electrons is energy-independent. Incidentally, the sign of the effect is the same when the carriers are holes rather than electrons, but there is a sign reversal if the electrons are scattered less strongly with increasing energy.

By contrast, Fig. 2.11(b) shows what happens in an intrinsic semiconductor or semimetal. Here, the longitudinal flows of the electrons and holes are in opposite directions, but both types of carriers experience forces through the magnetic field in the same (downward) transverse direction. Of course, the transverse flow of electrons and holes can occur without any overall current in this direction.

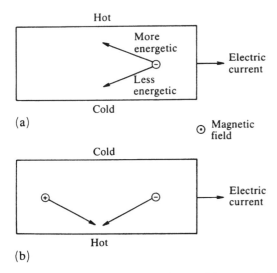

Fig. 2.11. The Ettingshausen effect in (a) an extrinsic semiconductor and (b) an intrinsic semiconductor or semimetal

Consequently, we can expect a relatively large Ettingshausen effect as the electrons and holes transport their excitative energy from one face of the sample to the other. When the magnetic field and current are in the specified directions, the top of the sample is cooled, and the bottom is heated.

The effects will be treated generally in terms of partial flows of carriers in bands 1 and 2, as before, but we are really interested only in the case where band 1 contains electrons and band 2 contains holes.

If the electric current density i_x is zero and there is no transverse temperature gradient, then from (2.78),

$$i_{1,x} = -i_{2,x} = \frac{\sigma_1 \sigma_2}{\sigma_1 + \sigma_2}(\alpha_2 - \alpha_1)\frac{\partial T}{\partial x}. \tag{2.103}$$

The action of the magnetic field B_z is to produce transverse partial currents $i_{1,y}$ and $i_{2,y}$ that must be equal and opposite if there is to be no overall current flow. This condition becomes established through the Nernst electric field $\mathcal{E}_y$. Thus,

$$i_{1,y} = \left(\mathcal{E} + R_{H,1} i_{1,x} B_z\right)\sigma_1, \qquad i_{2,y} = \left(\mathcal{E} + R_{H,2} i_{2,x} B_z\right)\sigma_2, \tag{2.104}$$

where $R_{H,1}$ and $R_{H,2}$ are the partial Hall coefficients, and

$$i_{1,y} + i_{2,y} = 0. \tag{2.105}$$

Eliminating the partial currents $i_{1,x}$ and $i_{2,x}$ by using (2.103), we find that the Nernst field is

$$\mathscr{E} = \frac{(R_{H,2}\sigma_2 - R_{H,1}\sigma_1)\sigma_1\sigma_2}{(\sigma_1 + \sigma_2)^2}(\alpha_2 - \alpha_1)B_z \frac{\partial T}{\partial x}, \qquad (2.106)$$

so that the bipolar Nernst coefficient is

$$N = \frac{(R_{H,2}\sigma_2 - R_{H,1}\sigma_1)\sigma_1\sigma_2}{(\sigma_1 + \sigma_2)^2}(\alpha_2 - \alpha_1). \qquad (2.107)$$

Now the Hall mobilities $\mu_{H,1}$ and $\mu_{H,2}$ are defined by the relationship

$$\mu_{H,1,2} = |R_{H,1,2}\sigma_{1,2}|. \qquad (2.108)$$

Thus, if $R_{H,2}$ is positive and $R_{H,1}$ is negative, as is the case when the carriers are holes and electrons, respectively,

$$N = \frac{(\mu_{H,2} + \mu_{H,1})\sigma_1\sigma_2}{(\sigma_1 + \sigma_2)^2}(\alpha_2 - \alpha_1). \qquad (2.109)$$

Note that, as $B_z \to \infty$, the Hall mobilities, $\mu_{H,1}$ and $\mu_{H,2}$, become equal to the mobilities, μ_1 and μ_2, respectively, defined by (2.62).

It is easy to express the electrical conductivity in terms of σ_1 and σ_2. Thus,

$$\sigma = \frac{\sigma_1}{1 + R_{H,1}^2 B_z^2 \sigma_1^2} + \frac{\sigma_2}{1 + R_{H,2}^2 B_z^2 \sigma_2^2}. \qquad (2.110)$$

However, conductivity is defined as the ratio of the current density in the x direction to the electric field in the same direction, when there is no transverse electric field. It is related to the isothermal electrical resistivity ρ^i by the expression [2.16]

$$\sigma = \frac{\rho^i}{(\rho^i)^2 + R_H^2 B_z^2}, \qquad (2.111)$$

where R_H is the overall Hall coefficient. At the high-field limit, $B_z \to \infty$, the value of R_H approaches $(1/R_{H,2} - 1/R_{H,1})^{-1}$.

As pointed out by Tsidil'kovskii [2.22], the general expression for the electronic thermal conductivity of a mixed semiconductor in an arbitrary magnetic field is very complex. However, bearing in mind that the lattice conductivity will undoubtedly predominate, an approximate expression for the electronic component should suffice. The simplest case is that when the transverse electric field is zero. Then a reasonable approximation at moderate to high magnetic field strengths is

$$\lambda_e = \frac{\lambda_{e,1}}{1+\mu_{H,1}^2 B_z^2} + \frac{\lambda_{e,2}}{1+\mu_{H,2}^2 B_z^2} + \frac{\sigma_1 \sigma_2 (\alpha_2 - \alpha_1)^2 T}{\sigma_1 (1+\mu_{H,1}^2 B_z^2) + \sigma_2 (1+\mu_{H,2}^2 B_z^2)}, \quad (2.112)$$

where $\lambda_{e,1}$ and $\lambda_{e,2}$ are the partial thermal conductivities in a zero magnetic field. This is consistent with Tsidil'kovskii's expression at the high-field limit, when $\mu_{H,1}^2 B_z^2 \gg $ and $\mu_{H,2}^2 B_z^2 \gg 1$.

It will be seen that λ_e, for $\mathcal{E}_y = 0$, tends to zero as the magnetic field becomes very large. However Tsidil'kovskii and others [2.24] have pointed out that, under the more usual boundary condition $i_y = 0$, λ_e of an intrinsic semiconductor tends toward a finite nonzero limit. This limit has the value $\lambda_L z_E T$, so that the total thermal conductivity becomes

$$\lambda = \lambda_L (1 + z_E T). \quad (2.113)$$

It is clear that the boundary conditions will be important for the materials in which we are especially interested.

2.6 Nonparabolic Bands

Now, we shall discuss a feature that may appear in the type of material that is used in thermoelectric and thermomagnetic refrigeration. In the foregoing parts of this chapter, it has been assumed that the bands are parabolic, that is, the energy of the carriers is proportional to $(k - k_0)^2$, where k_0 is the wave vector at the band extremum. This assumption is likely to become more or less invalid when the direct energy gap at the appropriate band edge is small and the carrier concentration is large. The problem is difficult, but a simplified approach has been presented by Kolodziejczak and Zhukotynski [2.25] and will be outlined here. The band will be assumed to have the form that was proposed by Kane [2.26] for the narrow-gap semiconductor, indium antimonide. The energy-wave-vector relationship in an isotropic case is then

$$E = \frac{\hbar^2 k^2}{2m} - \frac{1}{2}E_G + \frac{1}{2}\left(E_G^2 + \frac{8Q^2 k^2}{3}\right)^{1/2}, \quad (2.114)$$

where E_G is the direct gap. E_G may differ from the minimum gap that appears, for example, in the expression for the intrinsic carrier concentration. Q is a parameter that is characteristic of the particular band in question. The first term on the right-hand side of (2.114) may usually be neglected so that

$$E = \frac{1}{2}\left[\left(E_G^2 + \frac{8}{3}Q^2 k^2\right)^{1/2} - E_G\right]. \quad (2.115)$$

It will be seen that, when E_G is large compared with Qk,

$$E \approx \frac{2Q^2}{3E_G} k^2, \quad (2.116)$$

so that the band has the usual parabolic form close to the band edge. Therefore, we can identify $2Q^2/3E_G$ with $\hbar^2/2m^*$, where m^* is the effective mass at the edge of the band. When E_G is small compared with Qk,

$$E \approx \left(\frac{2}{3}\right)^{1/2} Qk, \quad (2.117)$$

and the relationship between E and k is linear. Thus, for high energies, d^2E/dk^2 tends to zero, and dE/dk is smaller than that for a parabolic band, so that, by any criterion that one might use, the effective mass becomes greater than m^*.

Kolodziejczak and Zhukotynski [2.25] define a "momentum" effective mass m_K from the relationship

$$\frac{1}{m_K} = \frac{1}{\hbar^2 k}\frac{dE}{dk}. \quad (2.118)$$

They also define a mobility $\mu_K = e\tau_e/m_K$ that differs from the mobility μ for a parabolic band. The transport integrals, defined by (2.94) and (2.95), can no longer be employed, and they must be written as

$$\mathcal{K}_s = -\frac{T}{3\pi^2 e}\int_0^\infty E^s \mu_K k^3 \frac{\partial f_0(E)}{\partial E}dE \Big/ (1 + \mu_K^2 B_z^2), \quad (2.119)$$

and

$$\mathcal{H} = -\frac{B_z T}{3\pi^2 e} \int_0^\infty E^s \mu_K^2 k^3 \frac{\partial f_0(E)}{\partial E} dE \bigg/ (1 + \mu_K^2 B_z^2), \quad (2.120)$$

where the density of states function is such that there are $k^2 dk/\pi^2$ states between k and $k + dk$. New transport integrals are defined by

$$L(\phi) = -\int_0^\infty \frac{\partial f_0(E)}{\partial E} k^3 \phi(E, \mu_K, B_z) dE, \quad (2.121)$$

where ϕ is a function of the energy, mobility, and magnetic field strength. Then,

$$\mathcal{H} = \frac{T}{3\pi^2 e} L\left(\frac{E^s \mu_K}{1 + \mu_K^2 B_z^2}\right), \quad (2.122)$$

and

$$\mathcal{H} = \frac{B_z T}{3\pi^2 e} L\left(\frac{E^s \mu_K^2}{1 + \mu_K^2 B_z^2}\right), \quad (2.123)$$

The various transport coefficients may be expressed in terms of the transport integrals L. Thus, the carrier concentration is

$$n = \frac{1}{3\pi^2} L(1), \quad (2.124)$$

and the electrical conductivity is

$$\sigma = \frac{e}{3\pi^2} L(\mu_K). \quad (2.125)$$

Now, if we wish to define a mobility as σ/ne, it has the value $L(\mu_K)/L(1)$. Other coefficients are

$$R_H = -\frac{3\pi^2 L(\mu_K^2)}{e L^2(\mu_K)}, \quad (2.126)$$

$$\alpha = -\frac{k_B}{e}\left[\frac{L(E\mu_K/k_B T)}{L(\mu_K)} - \eta\right], \tag{2.127}$$

and

$$N = \frac{k_B}{e}\frac{L(E\mu_K/k_B T)L(\mu_K^2) - L(\mu_K)L(E\mu_K^2/k_B T)}{L^2(\mu_K)}, \tag{2.128}$$

where it has been assumed that the carriers are electrons and the magnetic field approaches zero. In a very large magnetic field,

$$R_H = -\frac{1}{ne}, \tag{2.129}$$

and

$$\alpha = -\frac{k_B}{e}\left[\frac{L(E/k_B T)}{L(1)} - \eta\right]. \tag{2.130}$$

This illustrates the general principle that, as $B_z \to \infty$, the transport coefficients are independent of the energy dependence of the relaxation time.

It is expected that the nonparabolicity of the band will become significant only when the Fermi energy is much greater than $k_B T$, that is, when the electron gas is degenerate. If this is so, the Sommerfeld expansions may be used to give

$$L(\phi) = (\phi k^3)_\zeta + \frac{\pi^2}{6}(k_B T)^2\left[\frac{\partial^2(\phi k^3)}{\partial E^2}\right]_\zeta + ..., \tag{2.131}$$

$$L(E\phi/k_B T) = \eta(\phi k^3)_\zeta + \frac{\pi}{3}(k_B T)\left[\frac{\partial(\phi k^3)}{\partial E}\right]_\zeta + \frac{\pi^2}{6}(k_B T)^2\left[\frac{\partial^2(\phi k^3)}{\partial E^2}\right]_\zeta + ..., \tag{2.132}$$

where the subscript ζ indicates values of the Fermi energy. It is necessary to express μ_K as a function of k or E, and one can use the procedure described by Harman and Honig [2.27] for this purpose. They proposed that

$$\tau_e = \tau_1(k^2)^{r-1/2}\frac{dE}{dk}, \tag{2.133}$$

where τ_1 is a constant. This is a functional relationship that certainly holds for deformation-potential scattering and ionized-impurity scattering. Then,

$$\mu_K = \mu_0 k^{2r-3}\left(\frac{dE}{dk}\right)^2, \qquad (2.134)$$

where μ_0 is another constant. Thus, it is possible to set down reasonably concise expressions for the different coefficients. For example, the zero-field Seebeck coefficient is

$$\alpha = \mp \frac{2\pi^{2/3}}{3\hbar^2}\frac{k_B^2 T}{e}\frac{(m_K)_\varsigma}{(3n)^{2/3}}\left(r+\frac{3}{2}-3\frac{n}{m_K}\frac{\partial m_K}{\partial n}\right). \qquad (2.135)$$

The reader is referred to the works of Kolodzieczak and Zhukotynski [2.25] and of Harman and Honig [2.27] for further details of the technique.

2.7 Phonon Drag

Until now, it has been assumed implicitly that the flows of the charge carriers and of the phonons can be treated separately. However, it is often found that the interdependence of the flows must be taken into account, particularly at low temperatures. The resultant phenomena are known as the phonon-drag effects.

Phonon drag was first observed in measurements of the Seebeck coefficient of germanium at temperatures below about 100 K [2.28, 2.29]. The values that were obtained were much larger than could be accounted for by the theory outlined in Sect. 2.4, but Herring [2.30] showed that they could be explained by assuming that the electrons scattered the phonons preferentially in the direction of the current flow. The possibility of such an effect had been proposed previously [2.31, 2.32].

Although it is normally the phonon-drag Seebeck coefficient that is observed, it is conceptually easier to discuss the corresponding Peltier effect. The Kelvin relationship between the two coefficients still holds when phonon drag occurs.

Suppose that the concentration n of the charge carriers is very small. When one applies an electric field $\mathcal{E}$, these carriers take up momentum at the rate $\mp ne\mathcal{E}$ per unit volume. This momentum may be passed on to impurities and other defects where it is lost to random thermal vibrations. Alternatively, it may be passed on to phonons and retained until non-momentum-conserving collisions occur. If x is the fraction of collisions of the charge carriers that involve phonons and if the relaxation time for the loss of momentum from the phonon system is τ_d,

56   2. Transport of Heat and Electricity in Solids

$$\Delta p = \mp xne\tau_d \mathcal{E}, \tag{2.136}$$

where Δp is the excess momentum per unit volume carried by the phonons. It is important to realize that the time τ_d can be much greater than the relaxation time that is observed in the thermal conduction process because the electrons in a semiconductor are scattered predominantly by phonons of very low energy which undergo collisions far less frequently than the high-energy, heat-conduction phonons.

Because the electric current density is given by

$$I = ne\mu\mathcal{E}, \tag{2.137}$$

and the rate of heat flow per unit cross-sectional area is

$$w = v^2 \Delta p, \tag{2.138}$$

it follows that the phonon-drag Peltier coefficient is

$$\Pi_d = \frac{w}{i} = \mp\frac{xv^2\tau_d}{\mu}, \tag{2.139}$$

and the phonon-drag Seebeck coefficient is

$$\alpha_d = \frac{\Pi_d}{T} = \mp\frac{xv^2\tau_d}{\mu T}. \tag{2.140}$$

The phonon-drag thermoelectric coefficients reinforce the usual coefficients because both take the same sign as that of the charge carriers.

The phonon-drag effects should strongly depend on temperature. Typically τ_d varies as T^{-5} and μ as $T^{-3/2}$. Then α_d should vary as $T^{-9/2}$. Thus, it is unusual to observe phonon drag except at low temperatures, although it has been seen in semiconducting diamond above room temperature [2.33].

Note that (2.140) does not involve the carrier concentration but, in fact, $|\alpha_d|$ becomes smaller as the densities of carriers increase. This is associated to some extent with the scattering of the phonons by the impurities that are added to the semiconductor as doping agents. A more important effect is the transfer of momentum from the phonons back to the electrons, when the latter are very numerous. Herring [2.30] showed that this so-called saturation effect modifies (2.140), so that

$$\alpha_d = \mp \left(\frac{\mu T}{xv^2 \tau_d} + \frac{3nexv^2 \tau_d}{N_d k_B \mu T} \right)^{-1}, \tag{2.141}$$

where n is the carrier concentration and N_d the number of phonon modes that interact with the charge carriers. This dependence of phonon drag on carrier density makes it unlikely that the effect will ever be used to obtain worthwhile values of the figure of merit.

3. Selection and Optimization Criteria

3.1 Selection Criteria for Thermoelectric Materials

Perhaps more than any other, this chapter illustrates the challenge in obtaining materials with superior thermoelectric properties. A thorough understanding of solid-state physics and chemistry is required to define the potential of a particular system. The incorporation of different transport phenomena in evaluating the underlying transport mechanisms is illustrated.

In this chapter, we outline the basic physics necessary to understand and evaluate the transport properties of present and potential thermoelectric materials. The influence of one physical property on another will be apparent, and simple examples illustrate the basic principles. In Part II of the text we will build on these basic principles and outline an approach that has become the basis of much of materials research during the past decade. Chapter 6 and much of the work presented in Chap. 7 are a direct result of this novel approach.

3.2 Influence of Carrier Concentration on the Properties of Semiconductors

3.2.1 Thermoelectric Properties

It is not sufficient to nominate a particular semiconducting element or compound if a high z is required. It is necessary to specify the carrier concentration which can, of course, be adjusted by changing the number of donor or acceptor impurities. In this section, we discuss the problem of achieving the optimum concentrations of charge carriers.

It is instructional to proceed first with a calculation in which it is assumed that the semiconductor obeys classical statistics. We also suppose that there is only one type of carrier in a parabolic band, and we ignore the possibility of phonon drag. Thus, the thermoelectric parameters can be expressed by (2.62)–(2.67). Using the reduced Fermi energy η as the independent variable, we find that

$$zT = [\eta - (r + 5/2)]^2 / [(\beta \exp \eta)^{-1} + (r + 5/2)], \tag{3.1}$$

where the dimensionless materials parameter β, which was first introduced by Chasmar and Stratton [3.1], is defined by

$$\beta = \left(\frac{k_B}{e}\right)^2 \frac{\sigma_0 T}{\lambda_L}. \tag{3.2}$$

The parameter σ_0 that involves the carrier mobility and the density of states is given by

$$\sigma_0 = 2e\mu \left(\frac{2\pi m^* k_B T}{h^2}\right)^{3/2}. \tag{3.3}$$

If we use SI units and, in particular, express μ in m² V⁻¹ s⁻¹ and λ_L in W m⁻¹ K⁻¹, the values of σ_0 and β are

$$\sigma_0 = 774\mu \left(\frac{m^*}{m}\right)^{3/2} T^{3/2} \Omega^{-1}\,\text{m}^{-1}, \tag{3.4}$$

and

$$\beta = 5.745 \times 10^{-6} \frac{\mu}{\lambda_L}\left(\frac{m^*}{m}\right)^{3/2} T^{5/2}. \tag{3.5}$$

To find the optimum value of the reduced Fermi energy η_{opt}, we set $d(zT)/d\eta = 0$. This condition is met when

$$\eta_{opt} + 2\left(r + \frac{5}{2}\right)\beta \exp \eta_{opt} = r + \frac{1}{2}. \tag{3.6}$$

Now it turns out that $\beta < 0.5$ for all known materials. Also, the smallest value of r that is commonly suggested is $-3/2$ [3.2]. Thus, from (3.6), we conclude that $\eta_{opt} > -1.3$. This, of course, means that the assumption of classical statistics is invalid, because nondegeneracy is not a good approximation unless $\eta < -2$. For a typical semiconductor, we set $r \cong -1/2$, and $\beta < 1$, whence η_{opt} becomes close to zero, if (3.6) is applied.

In spite of the fact that we have demonstrated that the starting assumption of nondegeneracy cannot be justified, the previous calculation is not completely worthless. It shows correctly, for example, that if the Fermi energy is optimized by suitably adjusting of the carrier concentration, then z depends only on the parameters β and, to a lesser extent, on r. Furthermore, the conclusion that the optimum position for the Fermi level is close to the band edge is confirmed if the correct Fermi–Dirac statistics are employed. When it is invalid to use either the fully degenerate or the

3.2 Influence of Carrier Concentration on the Properties of Semiconductors

classical approximations, the appropriate expressions are (2.54) to (2.57), and the transport integrals K_s are defined by (2.52).

It is convenient to express the thermoelectric coefficients directly in terms of Fermi–Dirac integrals. Thus, the Seebeck coefficient is

$$\alpha = \pm \frac{k_B}{e} \left[\eta - \frac{\left(r+\frac{5}{2}\right) F_{r+3/2}(\eta)}{\left(r+\frac{3}{2}\right) F_{r+1/2}(\eta)} \right], \tag{3.7}$$

and the Lorenz number is

$$L = \frac{\lambda_e}{\sigma T} = \left(\frac{k_B}{e}\right)^2 \left\{ \frac{\left(r+\frac{7}{2}\right) F_{r+5/2}(\eta)}{\left(r+\frac{3}{2}\right) F_{r+1/2}(\eta)} - \left[\frac{\left(r+\frac{5}{2}\right) F_{r+3/2}(\eta)}{\left(r+\frac{3}{2}\right) F_{r+1/2}(\eta)} \right]^2 \right\}, \tag{3.8}$$

Also, if we use σ_0, as defined by (3.3), the electrical conductivity is given by

$$\sigma = \sigma_0 \frac{F_{r+1/2}(\eta)}{\Gamma\left(r+\frac{3}{2}\right)}. \tag{3.9}$$

Note, however, that the mobility μ in the expression for σ_0 is the value that would be observed in the nondegenerate region.

It is useful to present (3.7)–(3.9) in graphical form for particular values of the scattering exponent r equal to $-1/2$, $1/2$, and $3/2$. The range of special interest covers reduced Fermi energies, that lie between -2 and 4, because the classical and degenerate approximations are reasonably accurate outside these limits. In Fig. 3.1(a), $(e/k_B)\alpha$, which may be regarded as the dimensionless Seebeck coefficient, is plotted against η. Likewise, Fig. 3.1(b) and 3.1(c) show plots, also against η, of the dimensionless Lorenz number $(e/k_B)^2 L$ and the dimensionless electrical conductivity σ/σ_0.

The dimensionless figure of merit zT can be expressed in the form

$$zT = \left(\frac{e\alpha}{k_B}\right)^2 \left[\frac{\sigma_0}{\sigma\beta} + \left(\frac{e}{k_B}\right)^2 L \right]^{-1}, \tag{3.10}$$

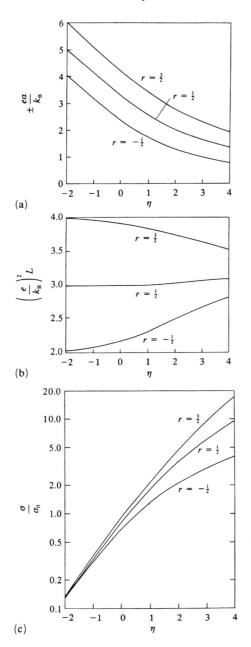

Fig. 3.1. Dimensionless thermoelectric coefficients plotted against reduced Fermi energy according to (3.7), (3.8), and (3.9); **(a)** Seebeck coefficient, **(b)** Lorenz number, **(c)** electrical conductivity

3.2 Influence of Carrier Concentration on the Properties of Semiconductors

which means that it is a function of only η, r, and β. Our task, then, is to find the optimum value of η and α, given the scattering parameter r and the materials parameter β. Then, the maximum z can be evaluated.

It is easy enough to determine the value of β if r is known and if the conductor is nondegenerate. Further, r can be found for a classical conductor (with the simplifying assumptions already made) from the temperature dependence of the carrier mobility. If the carriers are scattered by the thermal vibrations of the lattice, $\mu \propto T^{r-1}$, and if the scattering is due to static defects, $\mu \propto T^r$ [3.3]. The procedure is to find η from α, by using (2.65), and thence to obtain β from the figure of merit z, using (3.1). When the conductor is partially degenerate, the procedure is more complicated.

The problem has been tackled by Wasscher et al. [3.4], who showed how to determine $\beta \exp r$ from measurements of α, ρ, and λ of any extrinsic specimen, whether or not nondegeneracy is applicable. One first determines a value of the parameter β, which we shall call β^*, from (3.1) after assuming a particular value for r. Unless one has evidence to the contrary, r may be set equal to $-1/2$ the appropriate value for either acoustic-mode lattice scattering or alloy scattering. The true value of β can then be determined as the ratio β^*/f_w, where f_w is a function of α that has been calculated by Wasscher and his colleagues. Figure 3.2 shows the parameter f_w plotted against α, for $r = -1/2$, and also for $r = 1/2$ and $r = 3/2$.

After determing the true value of β, one can then find the optimum Seebeck coefficient α_{opt} at which the figure of merit has its maximum value z_{max}. The calculation of α_{opt} and $z_{max} T$ in terms of $\beta \exp r$ has been performed by Chasmar and Stratton [3.1]. Their results are presented graphically in Figs. 3.3 and 3.4. It is worth

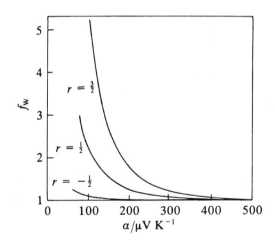

Fig. 3.2. The parameter f_w plotted against the Seebeck coefficient

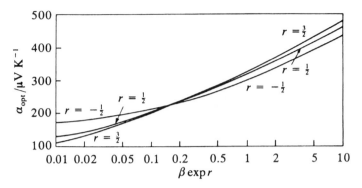

Fig. 3.3. Optimum Seebeck coefficient plotted against $\beta \exp r$

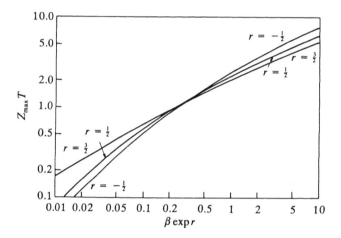

Fig. 3.4. Maximum figure of merit plotted against $\beta \exp r$

noting that all useful thermoelectric materials have values of $\beta \exp r$ between about 0.05 and 0.2, so that the optimum is invariably of the order of $\pm 200 \, \mu\mathrm{VK}^{-1}$.

From a practical point of view, the procedure is simplified by using the comprehensive curves that have been drawn up by Simon [3.5]. These curves allow one to predict α_{opt} from a single diagram and z_{max} using the data obtained on an arbitrary sample. Because any real material is unlikely to behave in an ideal manner, it is always necessary to optimize the properties empirically by measurements on a range of samples that have different carrier concentrations.

One of the factors that often has to be taken into account is the possible presence of minority carriers. In some instances, the predicted optimum Seebeck coefficient corresponds to a position for the Fermi level that is close enough to the opposite

3.2 Influence of Carrier Concentration on the Properties of Semiconductors

band edge so that contributions from both types of carriers are significant. It is quite clear that mixed conduction reduces z because of the opposite signs of the partial Seebeck coefficients and because of bipolar thermal conduction. Thus, the effect of a narrow energy gap E_g is to shift $|\alpha_{opt}|$ to a lower value.

To illustrate the effect of mixed conduction on z, we assume that there are two bands, 1 and 2, that contain majority and minority carrier, respectively, and contributions to the electrical conductivity are σ_1 and σ_2, such that $\sigma_1 \gg \sigma_2$. This means either that the carrier concentration n_1 is much greater than n_2 or that the mobility μ_1 is much greater than μ_2. Because we are looking for the relatively small perturbation from the z that would apply if there were no minority carriers, it will suffice to use the classical expression for partial thermoelectric coefficients. We shall also assume that the same scattering exponent r applies for both carrier types. Then,

$$|\alpha| = |\alpha_1| - \frac{k_B \sigma_2}{e \sigma_1}\left(2r + 5 + \frac{E_g}{k_B T}\right), \qquad (3.11)$$

$$\sigma = \sigma_1\left(1 + \frac{\sigma_2}{\sigma_1}\right), \qquad (3.12)$$

and

$$\lambda = \lambda_L + \sigma_1\left(\frac{k_B}{e}\right)^2 T\left[\left(r+\frac{5}{2}\right)\left(1+\frac{\sigma_2}{\sigma_1}\right) + \frac{\sigma_2}{\sigma_1}\left(2r+5+\frac{E_g}{k_B T}\right)^2\right], \qquad (3.13)$$

where α_1 is the partial Seebeck coefficient of the majority carriers. The ratio of the partial conductivities is given by

$$\frac{\sigma_2}{\sigma_1} = \frac{\mu_2\left(m_2^*\right)^{3/2}}{\mu_1\left(m_1^*\right)^{3/2}}\exp\left(-\frac{E_g}{k_B T} + 2\frac{e}{k_B}|\alpha_1| - 2r - 5\right). \qquad (3.14)$$

Because we are interested in the decrease from bipolar effects, we differentiate the expression for z with respect to the conductivity ratio to obtain

$$\frac{1}{z}\frac{\partial z}{\partial(\sigma_2/\sigma_1)} = \left[\frac{2}{\alpha}\left(\frac{\partial \alpha}{\partial(\sigma_2/\sigma_1)}\right) + \frac{1}{\sigma}\frac{\partial \sigma}{\partial(\sigma_2/\sigma_1)} - \frac{1}{\lambda}\frac{\partial \lambda}{\partial(\sigma_2/\sigma_1)}\right]. \qquad (3.15)$$

Thus, when $\sigma_1 \gg \sigma_2$,

$$\frac{\Delta z}{z_1} = \frac{\sigma_2}{\sigma_1}\left\{1 - \frac{2k_B}{e|\alpha_1|}\left(2r+5+\frac{E_g}{k_BT}\right) - \left(\frac{k_B}{e}\right)^2 \frac{z_1 T}{\alpha_1^2}\left[r+\frac{5}{2}+\left(2r+5+\frac{E_g}{k_BT}\right)^2\right]\right\}, \tag{3.16}$$

where z_1 is the figure of merit for $\sigma_2 = 0$.

To get some idea of what might happen in practice, we set $r = -1/2$ and assume that $z_1 T$ is approximately equal to unity, as it is for the best present-day materials. Then we find that

$$\frac{\Delta z}{z_1} \cong \frac{\sigma_2}{\sigma_1}\left\{1 - \frac{2k_B}{e|\alpha_1|}\left(4+\frac{E_g}{k_BT}\right) - \left(\frac{k_B}{e\alpha_1}\right)^2\left[2+\left(4+\frac{E_g}{k_BT}\right)^2\right]\right\}. \tag{3.17}$$

Furthermore, the optimized value of $|\alpha_1|$ is close to $3(k_B/e)$, so that if $\mu_1 \approx \mu_2$,

$$\frac{\Delta z}{(z_1)_{max}} \cong -\left[\frac{17}{9}+\frac{2E_g}{3k_BT}+\left(\frac{4}{3}+\frac{E_g}{3k_BT}\right)^2\right]\exp\left(-\frac{E_g}{k_BT}+2\right). \tag{3.18}$$

This equation shows that, if $E_g > 8k_BT$, $-\Delta z/(z_1)_{max} < 5\%$, so that, when the energy gap is large enough to satisfy this condition, minority-carrier conduction is unimportant. On the other hand, if $E_g \cong 6k_BT$, $-\Delta z/(z_1)_{max} \cong 30\%$, unless the Fermi level is shifted from the optimum position calculated for one type of carrier. In fact, the principal effect of a relatively small E_g is to increase the optimum Fermi energy and thus, reduce the partial Seebeck coefficient for the majority carrier. For example, we find that, if $|\alpha_1|$ is reduced from about $3(k_B/e)$ to $2(k_B/e)$, then the reduction of z below $(z_1)_{max}$, is no more than 14% for an energy gap of $6k_BT$.

Finally, note that it is quite possible for even an intrinsic semiconductor or semimetal to have useful thermoelectric properties if the ratio of the mobilities of the two types of carriers is large enough. Remembering that the right-hand side of (3.17) is dominated by the factor σ_2/σ_1 we shall assume that the densities of states in the two bands are sufficiently close for the Fermi level to lie near the center of the forbidden gap, in which case

$$|\alpha_1| = |\alpha_2| = \frac{k_B}{2e}\left(4+\frac{E_g}{k_BT}\right), \tag{3.19}$$

and

3.2 Influence of Carrier Concentration on the Properties of Semiconductors

$$\frac{\Delta z}{z_1} \cong -\frac{\mu_2}{\mu_1}\left[7+8\left(4+\frac{E_g}{k_B T}\right)^{-2}\right]. \tag{3.20}$$

Because the term in (3.20) that involves E_g is usually negligible, we see that $-\Delta z/z_1$, amounts to only about 7%, if $\mu_1/\mu_2 = 100$. There are indeed semiconductors, such as indium antimonide, in which the mobility ratio approaches this value [3.6].

3.2.2 Thermomagnetic Properties

It is quite easy to show that an extrinsic conductor will never have a thermomagnetic figure of merit z_E that is as high as its thermoelectric figure of merit z. Thus, Delves [3.7] demonstrated that (2.101) for the Nernst coefficient can be written approximately as

$$N \cong \frac{\mu}{eT\left(1+\mu^2 B_z^2\right)}\left[\frac{\langle \tau_e^2 E \rangle}{\langle \tau_e^2 \rangle} - \frac{\langle \tau_e E \rangle}{\langle \tau_e \rangle}\right], \tag{3.21}$$

where τ_e is the electronic relaxation time and the angular brackets indicate average values taken over all energies. The factor in the square brackets lies between zero for fully degenerate material and $rk_B T$ for a completely nondegenerate conductor. We see that the so-called thermomagnetic power NB_z becomes equal to zero when $\mu B_z \gg 1$ and reaches a maximum value of $rk_B/2e$ when $\mu B_z \cong 1$. Because $|r|$ is unlikely to be greater than unity, the thermomagnetic power will probably always be less than $k_B/2e$, or about 43 µV K^{-1}. Now, we have already shown that the optimized Seebeck coefficient for a thermoelectric material is of the order of 200 µV K^{-1}. This means that the value of z_E will generally be very much less than z when there is only one type of carrier.

Although bipolar effects are deleterious to the thermoelectric z, we shall see that they can lead to high values of z_E. We have already discussed this qualitatively in Chap. 2, with reference to Fig. 2.11. It will be assumed that the magnetic field is high enough, so that both $(\mu_1 B_z)^2$ and $(\mu_2 B_z)^2$ are much greater than unity and that the electron and hole concentrations are equal, that is, the conductor is fully intrinsic. The latter condition ensures that the thermomagnetic power will increase without limit, as the magnetic field rises.

From (2.109) for an intrinsic conductor,

$$N = \frac{\mu_1 \mu_2}{\mu_1 + \mu_2}(\alpha_2 - \alpha_1). \tag{3.22}$$

In a high magnetic field, the Hall coefficient R_H, also becomes equal to zero for equal electron and hole concentrations. Thus, from (2.110) and (2.111),

$$\rho^i = \frac{1}{\sigma} = \left(\frac{\sigma_1(0)}{1+\mu_1^2 B_z^2} + \frac{\sigma_2(0)}{1+\mu_2^2 B_z^2} \right)^{-1}, \qquad (3.23)$$

where $\sigma_1(0)$ and $\sigma_2(0)$ are the partial conductivities in a zero magnetic field. Then, if we express σ_1 as $n_i e \mu_1$ and σ_2 as $n_i e \mu_2$, and set $\mu_1^2 B_z^2 \gg 1 \ll \mu_2^2 B_z^2$,

$$\rho^i = \frac{\mu_1 \mu_2}{(\mu_1 + \mu_2) n_i e} B_z^2. \qquad (3.24)$$

Furthermore, the isothermal thermal conductivity, as defined in Chap. 1, tends to the limiting value λ_L, as the magnetic field becomes large. Thus, z in a high magnetic field tends toward

$$z_E = \frac{n_i e \mu_1 \mu_2 (\alpha_2 - \alpha_1)}{(\mu_1 + \mu_2) \lambda_L}. \qquad (3.25)$$

In a sense, we cannot optimize the carrier concentration in an intrinsic conductor for thermomagnetic cooling in the same way that we can for an extrinsic thermoelectric material. The intrinsic carrier density n_i is a fixed quantity at a given temperature. However, one can use (3.25) to determine the preferred value of E_g for given values of the other relevant parameters. Both n_i and $(\alpha_2 - \alpha_1)$ depend on E_g.

Let us first suppose that classical statistics might be applicable. Then,

$$n_i = 2(m_1^* m_2^*)^{3/4} \left(\frac{2\pi m k_B T}{h^2} \right)^{3/2} \exp\left(-\frac{E_g}{2k_B T} \right), \qquad (3.26)$$

and, at the high-field limit,

$$(\alpha_2 - \alpha_1) = \frac{E_g + 5k_B T}{eT}. \qquad (3.27)$$

Thus,

$$z_E \propto \left(E_g + 5k_B T \right)^2 \exp\left(-\frac{E_g}{2k_B T} \right). \qquad (3.28)$$

3.2 Influence of Carrier Concentration on the Properties of Semiconductors

We find that $dz_E/dE_g = 0$ when $E_g = -k_B T$. Therefore, this appears to be the optimum band gap. In other words, it is predicted that the material is not a semiconductor but a semimetal in which the conduction and valence bands overlap by $k_B T$. This, of course, again invalidates the use of classical statistics.

To demonstrate the effect of using Fermi–Dirac statistics, we shall assume that the effective masses m_1^* and m_2^* are approximately equal, so that the Fermi level lies halfway between the band edges. Then, it can be shown that

$$z_E \propto F_{r+1/2}(\eta) \left[\frac{5F_{3/2}(\eta)}{3F_{1/2}(\eta)} - \eta \right]^2, \tag{3.29}$$

where η, the reduced Fermi energy, is equal to $-E_g/2k_B T$. The variation of the high-field z_E, normalized to its value at zero E_g, is plotted as a function of the energy gap in Fig. 3.5. It may be seen that, with the usually assumed value of $-1/2$ for the scattering parameter r, z_E has a rather flat maximum at a band overlap of about $2k_B T$. The optimum band overlap increases with r and would occur at strong degeneracy for $r = 3/2$. However, it must be remembered that such a value of r would

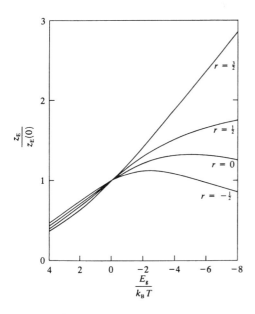

Fig. 3.5. Ratio of thermomagnetic figure of merit to its value at zero energy gap, plotted as a function of the reduced energy gap

correspond to ionized-impurity scattering which would make it extremely difficult to achieve the high magnetic field condition, and, in any case, such scattering is most unlikely in an intrinsic conductor. Thus, there is every indication that the preferred condition is a small band overlap, although a semiconductor with a very small energy gap must not be ruled out as a potential thermomagnetic material.

For a given value of the energy gap, the high-field z_E from (3.25) satisfies the proportionality

$$z_E \propto \left(m_1^* m_2^*\right)^{3/4} \left[\left(\frac{1}{\mu_1} + \frac{1}{\mu_2}\right)\lambda_L\right]^{-1}. \tag{3.30}$$

Thus, the materials factor $(\mu/\lambda_L)(m^*/m)^{3/2}$ that appears in the parameter β for a thermoelectric material is replaced by

$$\left(m_1^* m_2^*\right)^{3/4} \left[(1/\mu_1 + 1/\mu_2)\lambda_L\right]^{-1} \tag{3.31}$$

in the thermomagnetic case. We notice that, if $m_1^* = m_2^*$ and $\mu_1 = \mu_2$, the two factors become equivalent. It should also be observed that if one of the carriers has a much smaller mobility than the other, say, $\mu_2 \ll \mu_1$, the materials factor in the proportionality (3.30) tends toward $\mu_2(m_1^* m_2^*)^{3/4}/\lambda_L$, that is, the mobility of the less mobile carrier is the only one that matters. Therefore, it is important, that the mobilities of both carriers should be as high as possible in a good thermomagnetic material. It goes without saying that high carrier mobilities make it easier to reach the high-field condition with readily available magnetic fields.

If $\mu_1 = \mu_2 = \mu$, (3.25) becomes

$$z_E = \frac{2 n_i e \mu \alpha^2}{\lambda_L}, \tag{3.32}$$

where we have set $\alpha_2 = -\alpha_1 = \alpha$. Thus, we may compare the z_E of a thermomagnetic refrigerator and that of a thermoelectric refrigerator in which electrons and holes that have identical properties occupy the two branches of each sample. The thermomagnetic refrigerator has the following advantages:

(1) The factor of 2 in the numerator, which appears because the two types of charge carrier share a common lattice.
(2) The thermal conductivity contains only the lattice contribution.
(3) If r is negative, the value of α is larger than it would be in a zero magnetic field.

Of course, it would be possible to achieve advantages (2) and (3) for a thermoelectric refrigerator by operating it in a high magnetic field but advantage (1) is inherent in the thermomagnetic device.

3.3 Optimization of Electronic Properties

3.3.1 Simple Bands

In Sect. 3.2, we showed that the principal requirement for charge carriers is that $\mu(m^*)^{3/2}$ should be as high as possible. There is the subsidiary requirement for a thermoelectric material that E_g should be large enough to prevent minority-carrier conduction from playing too great a role. On the other hand, for thermomagnetic refrigeration, the energy gap must be small enough, so that there are large concentrations of both electrons and holes. In the latter case, $\mu(m^*)^{3/2}$ should be large for both types of carrier, and also the mobilities themselves should be high enough for μB_z to exceed unity in a practically attainable magnetic field. In this context, then, we discuss selecting of semiconductors and semimetals that have favorable electronic properties.

Let us suppose, for the moment, that the material is isotropic and has a single-valley conduction or valence band centered at $k = 0$. This allows us to discuss the effective mass as a scalar quantity. At first sight, it might appear that a large effective mass is desirable, but this is not necessarily true, because mobility is related to the effective mass, μ usually increases as m^* decreases. The precise form of the interdependence between μ and m^* is governed by carrier-scattering processes.

In pure crystals, carriers are scattered by thermal vibrations of the crystal lattice. The simplest type of lattice scattering occurs when there is only one atom per unit cell; then the vibrations are entirely of the acoustic type. Actually, acoustic-mode lattice scattering also predominates in many crystals that have more than one atom per unit cell. The scattering results from fluctuations in the periodic potential, that are brought about by local density changes as the interatomic spacing changes. Bardeen and Shockley [3.8] first succeeded in quantifying this scattering process using the so-called deformation-potential method. They obtained the following expression for relaxation time:

$$\tau_e = \frac{h^4 v^2 \rho_d E^{-1/2}}{\left(8\pi^2\right)^{3/2} k_B T \Delta^2 \left(m^*\right)^{3/2}}, \tag{3.33}$$

where ρ_d is the density and Δ is the deformation-potential constant. Δ corresponds to the volume dependence of the potential energy at the band edge. It will be observed that the parameter $r = -1/2$ for acoustic-mode lattice scattering. Thus, from (2.64) and (3.33),

$$\mu = \frac{\left(8\pi\right)^{1/2} e\hbar v^2 \rho_d}{3 k_B^{3/2} \Delta^2} \left(m^*\right)^{-5/2} T^{-3/2}. \tag{3.34}$$

Therefore, $\mu(m^*)^{3/2}$ should be proportional to $(m^*)^{-1}$, and the effective mass should be small rather than large.

Sometimes the mobility falls more rapidly with increasing temperature, than would be indicated by (3.34). For example, in lead chalcogenides, μ is approximately proportional to $T^{-5/2}$, suggesting that r = -3/2. This is not consistent with any known scattering process involving single phonons, though it could be explained if each scattering event involved two phonons [3.9]. Of course, any temperature dependence of the effective mass would also have a considerable effect on the change of mobility with temperature.

When optical modes of vibration are present, they may compete with acoustic modes in scattering carriers. They, too, can give rise to deformation-potential scattering [3.10]. However, when crystal binding is essentially ionic, scattering is primarily caused by the polarization that results from periodic displacements. The strength of the interaction depends on a parameter α_i, given by

$$\alpha_i = \frac{e^2}{2h}\left(\frac{m^*}{2h\upsilon_0}\right)^{1/2}\frac{\varepsilon - \varepsilon'}{\varepsilon\varepsilon'}, \tag{3.35}$$

where υ_0 is the optical-mode characteristic frequency and ε and ε' are the static and high-frequency dielectric constants, respectively. The difference between ε and ε' becomes greater as the material becomes more strongly ionic, and this increases the strength of the interaction between the carriers and the optical phonons. When $\alpha_i > 1$, as it is in very strongly ionic crystals, the conduction process is discussed in terms of polaron theory [3.11]. In this case, each carrier transports the local polarization of the lattice, as it moves through the crystal. However, μ associated with such a process is very small, and it seems most unlikely that it is relevant in any potentially useful thermoelectric material. Likewise, we shall ignore the possibility of charge transport by the ions themselves. Ionic semiconductors would be unsuitable for thermoelectric energy conversion because of the chemical changes that occur whenever they carry a direct current [3.3].

Optical-mode scattering, for $\alpha_i < 1$, has been treated by perturbation theory [3.12]. Difficulties arise from the fact that the relaxation-time approximation cannot be used. A simplified expression for mobility is

$$\mu = \frac{4h^2\varepsilon e(k_BT)^{1/2}}{3\pi^{3/2}\alpha_i m^*(h\upsilon_0)^{3/2}}\chi\left[\exp\left(\frac{h\upsilon_0}{k_BT}\right)-1\right], \tag{3.36}$$

where χ is a slowly changing function that varies from 1 to 0.6 as $h\upsilon_0/k_BT$ takes values between 0 and 3. Because of the exponential term, μ could become very high at low temperatures according to (3.36), but this probably implies that some other form of scattering would then take over. At high temperatures, the relaxation time approach can also be used, leading to

$$\tau_e \propto \frac{E^{1/2}}{(m^*)^{1/2}T}, \qquad (3.37)$$

where μ varies as $(m^*)^{-3/2}T^{-1/2}$. Thus, when this type of polar scattering dominates, the scattering parameter r is equal to 1/2 and the product $\mu\,(m^*)^{3/2}$ should be independent of the effective mass.

According to Ehrenreich [3.13], one can actually use the relaxation-time approximation to determine the transport properties of a polar semiconductor to temperatures as low as $2h\upsilon_0/k_B$, with $r = 1/2$, and even down to a temperature equal to $h\upsilon_0/k_B$, provided that r is regarded as an adjustable parameter.

Turning to scattering by static imperfections, one might think that the ionized donor and acceptor impurities would have an important effect in limiting μ in view of the rather large carrier concentrations that are required in thermoelectric devices. The original treatment of ionized-impurity scattering was due to Conwell and Weisskopf [3.14] who found that the corresponding relaxation time is given by

$$\tau_e = \frac{8(2)^{1/2}(m^*)^{1/2}\varepsilon^2 E^{3/2}}{\pi^{3/2} N_i e^4 N}\left[1+\left(\frac{3\varepsilon k_B T}{e^2 N_i^{1/3}}\right)^2\right]^{-1}, \qquad (3.38)$$

where N_i is the impurity concentration. It turns out that ionized-impurity scattering is not usually too important in thermoelectric materials because they have rather high values of ε and at high impurity densities, there is a screening effect that considerably reduces the collision probability [3.15].

In fact, Ioffe [3.16] suggested that ionized-impurity scattering might be beneficial, even though it would lead to a smaller μ than that for lattice scattering alone. The point is that the scattering parameter r rises to 3/2 when ionized-impurity scattering is dominant and, as may be shown from (2.65), this substantially increases the value of α for a given carrier concentration. There is not any experimental evidence to support Ioffe's suggestion, and Ure [3.17] has calculated that, even using compensated impurities to maintain the Fermi energy at its optimum level, the improvement in Z could never amount to more than about 10%.

Scattering by neutral impurities [3.18] and dislocations [3.19] is unlikely to be important at ordinary temperatures, but alloy scattering is certainly likely to be a factor, because, as we shall see, the best thermoelectric materials are invariably solid solutions rather than pure elements or compounds. Brooks [3.15] showed that scattering in alloys can be treated by deformation-potential theory, because it results from local variations in potential energy at band edges. In particular, the scattering parameter r takes the same value, $-1/2$, as it has for acoustic-mode lattice scattering. However, because the imperfections are static, there is no temperature dependence of the relaxation time for carriers of a given energy, so that the mobility varies as $T^{-1/2}$. Thus, in semiconductor solid solutions, a change in the temperature dependence of the mobility does not necessarily imply a change in r.

3.3.2 More Complex Band Structures

So far it has been implied that the surfaces of constant energy in k space are spherical and centered at the origin. A parabolic relationship between the energy E and the wave vector k has also been assumed. In this section, more general band structures are considered.

If the crystal structure is noncubic, the surfaces of constant energy become ellipsoidal, even if they are at $k = 0$. Then the effective mass is different in different directions and must be regarded as a tensor rather than a scalar quantity. We may suppose that the values of the effective mass along the principal axes are m_1, m_2, and m_3. Then, if the relaxation time is isotropic, the mobilities in the three directions are proportional to $1/m_1$, $1/m_2$, and $1/m_3$, respectively.

When the energy extrema are found elsewhere than at the center of the Brillouin zone, the constant-energy surfaces are generally ellipsoidal, even in a cubic crystal. Let us consider a multivalley cubic crystal. Then, the transport properties of each valley will be anisotropic, though the overall transport coefficients, in the absence of a magnetic field, must be the same in all directions. It is convenient to define an inertial mass m_I that may be used to relate the mobility to the relaxation time, according to (2.64). This inertial mass is given by

$$\frac{1}{m_I} = \frac{1}{3}\left(\frac{1}{m_1} + \frac{1}{m_2} + \frac{1}{m_3}\right), \tag{3.39}$$

where m_1, m_2, and m_3 are the principal effective masses within each valley. On the other hand, the density of states in any valley, as expressed by (2.34), depends on the geometric mean of m_1, m_2, and m_3. Thus we define a single-valley, density-of-states mass m_N by

$$m_N = \left(m_1 m_2 m_3\right)^{1/3}. \tag{3.40}$$

Then, if there are N_v valleys, as required by the positions of the extrema and the symmetry of the crystal, the overall density-of-states effective mass (for which we retain the symbol m^*) is given by

$$m^* = N_v^{2/3} m_N = N_v^{2/3}\left(m_1 m_2 m_3\right)^{1/3}, \tag{3.41}$$

This overall density-of-states mass determines the carrier concentration that corresponds to a given value of the Fermi energy.

The previous equations suggest that it might be possible to combine the virtues of a high density-of-states mass m^* and a high μ by using a multivalley semiconductor. For example, in (3.34) for the relaxation time, the relevant effective mass becomes m_N, when acoustic-mode lattice scattering predominates. The mobility is

3.3 Optimization of Electronic Properties

proportional to $m_N^{-3/2} m_I^{-1}$, and $\mu(m^*)^{3/2} \propto N_v/m_I$. Thus, we require a large number of valleys, where each has a small inertial effective mass. On the other hand, when the semiconductor is multivalleyed, intervalley scattering becomes possible, and this may substantially reduce the carrier relaxation time [3.20].

Detailed calculations of the effect of intervalley scattering on the thermoelectric properties have been carried out by Rowe and Bhandari [3.22]. Their results apply specifically to lead telluride, but their general conclusion that it is still favorable to use a multivalley semiconductor, in spite of significant intervalley scattering, probably holds for other materials as well. It is certainly true that the best materials for thermoelectric applications at all temperatures have multivalley band structures.

Although a discussion in terms of acoustic-mode lattice scattering indicates the desirability of a small value for inertial mass m_I, we cannot be certain that extremely low values of this parameter are wanted. For example, (3.39) shows that a small value of the effective mass makes ionized-impurity scattering more probable. Perhaps we can say no more than that multivalley semiconductors, whose carriers have a rather small inertial mass are likely to be the best thermoelectric or thermomagnetic materials.

One cannot define a single inertial mass for a noncubic semiconductor and, for such a material, the mobility will be anisotropic. There will be a preferred direction for current flow, and this is likely to be the one for which the inertial mass has its smallest value. Even in such anisotropic materials, these considerations should be generally applicable.

Rowe and Bhandari [3.21] considered the effect of the nonparabolicity of energy bands on z. Their general conclusion is that nonparabolicity is most undesirable. Presumably, the increase in the density of states as the energy rises is more than offset by a decrease in mobility, compared with the value when the carriers are in states near the band edge. However, it may be that semiconductors that have small values for the direct energy gap and, thus, strongly nonparabolic bands are those that are most likely to have the highest carrier mobilities.

In the theory of nonparabolic bands in Sect. 2.6, we introduced the parameter Q that could be identified with $(\hbar/2)(3E_g m^*)^{1/2}$ in a single-valley conductor. It is interesting to note that Q has more or less the same value for the conduction band of all III–V and II–VI compounds [3.23] and even for the noncubic semiconductor Cd_3As_2 [3.24]. This gives some support for the link between high mobilities and small energy gaps mentioned before. Then, it seems probable, that we should be looking for semiconductors with small rather than large energy gaps, at least for applications at ordinary temperatures.

3.4 Minimizing Thermal Conductivity

3.4.1 Atomic Weight and Melting Point

First, we consider pure elements and compounds, and we direct our attention to the so-called high-temperature region, over which lattice conductivity varies as T^{-1}. This region usually extends below room temperature and, for most materials of interest, even below liquid nitrogen temperature. It may be assumed that phonon–phonon scattering is dominant under the conditions stated.

Some success in predicting high-temperature lattice conductivity was achieved by Leibfried and Schlömann [3.25]. By using the variational method, they found that

$$\lambda_\mathrm{L} = 3.5 \left(\frac{k_\mathrm{B}}{h}\right)^3 \frac{MV^{1/3}\theta_\mathrm{D}^3}{\gamma^2 T}, \qquad (3.42)$$

where M is the average mass per atom, V is the average atomic volume, and γ is the Grüneisen parameter, which enters into the equation of state for solids and, together with the expansion coefficient, is a measure of the anharmonicity of lattice vibrations.

One can obtain Leibfried and Schlömann's formula, or rather one that differs from it only in the value of the numerical constant, by the following simple means. We start by using Dugdale and MacDonald's argument that there should be a relationship between λ_L and the thermal expansion coefficient α_T, because both depend on the anharmonicity of interatomic forces. Dugdale and MacDonald [3.26] showed that anharmonicity can be represented by the dimensionless quantity $\alpha_\mathrm{T}\gamma T$, and they suggested that the phonon free path length l_t should be approximately equal to the lattice constant a divided by this quantity. Thus, by substitution in (2.7),

$$\lambda_\mathrm{L} = \frac{c_\mathrm{V} a v}{3\alpha_\mathrm{T}\gamma T}. \qquad (3.43)$$

We may eliminate the expansion coefficient from this equation by using the Debye equation of state

$$\alpha_\mathrm{T} = \frac{\chi\gamma c_\mathrm{V}}{3}, \qquad (3.44)$$

where χ is compressibility. The speed of sound v and χ are also related to the Debye temperature θ_D through the equation

$$v = (\rho_d \chi)^{-1/2} = \frac{2k_B a \theta_D}{h}, \tag{3.45}$$

where ρ_d is the density. Then, if a is set equal to the cube root of the atomic volume,

$$\lambda_L \approx 8 \left(\frac{k_B}{h} \right)^3 \frac{M V^{1/3} \theta_D^3}{\gamma^2 T}, \tag{3.46}$$

which may be compared with (3.42). Because we are not concerned with the value of the numerical constant here, we may use either (3.42) or (3.46), and we opt for the latter.

Neither (3.43) nor (3.46) is as useful as one might hope in predicting the value of λ_L. One of the relationships requires a knowledge of the expansion coefficient and the speed of sound, whereas the other involves 2_D. Keyes [3.27] showed that a more useful formula could be obtained by using further approximations. His starting point was Lawson's [3.28] equation for thermal conductivity,

$$\lambda_L \approx \frac{\alpha}{3\gamma^2 T \chi^{3/2} \rho_d^{1/2}}, \tag{3.47}$$

which is easily obtained from (3.43) or (3.46) by substituting of (3.45). Keyes then eliminated compressibility by using the Lindemann melting rule,

$$\chi \approx \frac{\varepsilon_m V}{R T_m}, \tag{3.48}$$

where R is the gas constant and T_m is the melting temperature. This rule is based on the assumption that a solid melts when the amplitude of the lattice vibrations reaches a fraction ε_m of the lattice constant; ε_m is approximately the same for all substances. Keyes found that

$$\lambda_L T \approx B \frac{T_m^{3/2} \rho_d^{2/3}}{A^{7/6}}, \tag{3.49}$$

where

$$B = \frac{R^{3/2}}{3\gamma^2 \varepsilon_m^3 N_A^{1/3}}, \tag{3.50}$$

where N_A is Avogadro's number and A is the mean atomic weight.

The factor B involves only the universal constants R and N_A and two quantities, γ and ε_m, that are unlikely to change much from one material to another. The variables T_m, ρ_d, and A in (3.49) for thermal conductivity are, of course, known as soon as a material is first synthesized, so that, if the equation is valid, it is a most useful aid for predicting λ_L.

By collecting data for a wide variety of dielectric crystals, Keyes showed, that (3.49) is always satisfied within an order of magnitude if B is set equal to 3×10^{-4} SI units. If one selects different values of B for covalent and ionic crystals, much better agreement is obtained. Thus, one may use B equal to 1.3×10^{-3} SI units for covalent materials and 1.5×10^{-4} SI units for ionic materials. Good agreement between the predictions of the Keyes rule, (3.49), and the experimental data for a variety of semiconductors using a value of 6×10^{-4} SI units for B, as shown in Fig. 3.6, has been found [3.29].

Note that elements or compounds of high mean atomic weight tend to have low melting temperatures. Their densities are also usually high, but we expect that $\rho_d^{2/3} A^{-7/6}$ should decrease as atomic weight rises. Thus, the Keyes rule supports the prediction that materials of high mean atomic weight should have low thermal conductivity and should be chosen for thermoelectric applications.

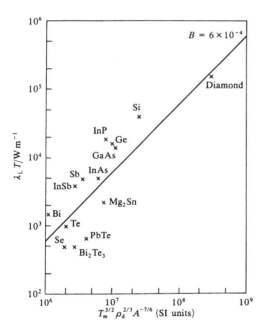

Fig. 3.6. Product of lattice conductivity and absolute temperature plotted against $T_m^{3/2} \rho_d^{2/3} A^{-7/6}$ for a number of semiconductors

Ioffe and Ioffe [3.31] also drew attention to the dependence of thermal conductivity on atomic weight and demonstrated a difference between covalent and ionic crystals that is consistent with the different values of B that are preferred for the respective types of material. Their results are illustrated in Fig. 3.7. One should not infer that ionic materials are likely to be better than covalent semiconductors in thermoelectric energy conversion on account of their lower range of thermal conductivity. Ionic compounds invariably have very much smaller carrier mobilities, and the factor $\mu(m^*)^{3/2}/\lambda_L$ is highest for semiconductors in which covalent bonding predominates.

3.4.2 Semiconductor Solid Solutions

Ioffe et al. [3.32] first made the interesting and useful suggestion that solid solutions or alloys might be superior to elements or compounds in thermoelectric properties. In forming a solid solution between isomorphous crystals, the long-range order would be preserved so that the charge carriers, that have comparatively long wavelengths would not suffer a reduction in mobility. On the other hand, the

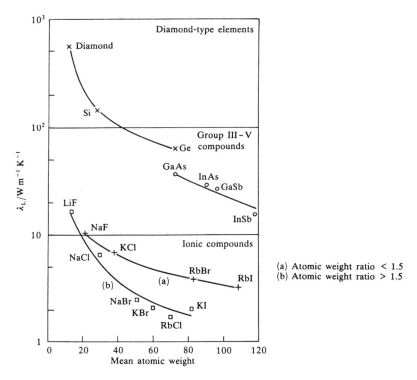

Fig. 3.7. Lattice conductivity at room temperature plotted against mean atomic weight for covalent ionic crystals

phonons that contribute most to the transport of heat have a much shorter wavelength and would be effectively scattered by the short-range disorder, thus reducing thermal conductivity.

Shortly afterward, Airapetyants et al. [3.33] proposed a refinement of this principle. They considered that electrons would be more strongly scattered by disturbances in the electropositive sublattice of a compound, whereas holes would be more strongly scattered by disturbances in the electronegative sublattice. On this basis, positive and negative thermoelectric materials should be produced from solid solutions in which there is disorder only in the electronegative and electropositive sublattices, respectively.

Although there is some experimental evidence to support these views, they cannot be regarded as firmly established. Indeed, they imply that there should be strong scattering of both electrons and holes in solid solutions between elemental semiconductors, whereas silicon–germanium alloys have proved most useful as high-temperature thermoelectric materials [3.34].

Whether account should be taken of the principle of Airapetyants and his colleagues, it is clear that the reduction of lattice thermal conductivity in certain solid solutions of both elements and compounds, outweighs any fall in μ of the appropriate charge carriers. Therefore, it is important to consider the mechanism by which phonon scattering occurs in such materials.

According to the Rayleigh theory, the scattering cross-section σ for point defects can be expressed as

$$\sigma = \frac{4\pi c^6 q_L^4}{9}\left(\frac{\Delta\chi}{\chi} + \frac{\Delta\rho_d}{\rho_d}\right)^2, \tag{3.51}$$

where c is the linear dimension of the defect, q_L is the magnitude of the phonon wave vector, $\Delta\chi$ is the local change of compressibility, and $\Delta\rho_d$ is the local density change. The applicability of Rayleigh scattering has been discussed by Klemens [3.35]. Strictly speaking, the dimensions of the defect should be much smaller than the wavelength of the radiation. For higher frequency phonons, however, this is not true, even if c is no larger than the interatomic spacing. Nevertheless, the use of the Rayleigh formula can be justified because the short-wavelength phonons are so strongly scattered by point defects that they contribute very little to thermal conductivity. Most of the heat transport then arises from phonons for which the Rayleigh theory is applicable. The same argument allows us to use the Debye model for lattice vibrations of alloys with more confidence than would be the case for simple elements or compounds.

Although, as shown in Chap. 2, we might expect normal processes to have some influence on thermal conductivity, even at high temperatures, we can obtain a good idea of the effect of alloy scattering by assuming that the umklapp processes are predominate ($k_0 \rightarrow 0$). Then, (2.25) becomes

3.4 Minimizing Thermal Conductivity

$$\frac{\lambda_L}{\lambda_0} = \frac{\omega_0}{\omega_D} \tan^{-1}\left(\frac{\omega_D}{\omega_0}\right). \tag{3.52}$$

where ω_0/ω_D is defined by (2.27). Here, λ_L is the thermal conductivity of the solid solution and λ_0 is that of the so-called virtual crystal; λ_0 is the lattice conductivity that the solid solution would possess if it were perfectly ordered, that is, if point-defect scattering were absent. The value of λ_0 can be obtained by linear interpolation between the observed thermal conductivities of the components of the solid solution, if no better estimate can be made. Note that, for group IV elements, elastic constants vary inversely with the fourth power of the lattice spacing; this enables estimating θ_D accurately for any particular alloy and certainly assists in predicting λ_0 and the speed of sound v.

Now, (3.51) shows that the phonons are scattered by variations of both elasticity and density. The density or mass-fluctuation effect is easier to determine theoretically and might, in fact, predominate in the type of alloy in which we are interested. In this case, the parameter A in (2.27) is given by

$$A_M = \frac{\pi}{2v^3 N} \sum_i \frac{x_i (M_i - \overline{M})^2}{\overline{M}^2}. \tag{3.53}$$

where x_i is the concentration of unit cells of mass M_i, $\overline{M}$ is the average mass per unit cell, and N is the number of unit cells per unit volume.

The additional effect of scattering due to elasticity fluctuations, or strain scattering, is more difficult to predict. However, we can outline the approach. A foreign atom changes local compressibility because it has bonds different from those of the host atoms and also because it does not fit properly into a lattice site, thus straining the crystal. Based on the elastic continuum model, an impurity atom of width δ_I' in its own lattice will distort the space it occupies from one having a width δ in the host lattice to one of width δ_I, where

$$\frac{\delta_I - \delta}{\delta} \equiv \frac{\Delta \delta_i}{\delta} = \frac{\mu}{(1+\mu)} \frac{\delta_I' - \delta}{\delta}, \tag{3.54}$$

and

$$\mu = \frac{(1+v_P)G_I}{2(1+2v_P)G}. \tag{3.55}$$

G and v_P are the bulk modulus and Poisson's ratio for the host crystal, respectively, and G_I is the bulk modulus of the impurity crystal. Klemens has shown that the value of A in (2.27) for strain scattering is given by

$$A_s = \frac{\pi}{v^3 N} \sum_i x_i \left(\frac{\Delta G_I}{G} - 6.4\gamma \frac{\Delta \delta_i}{\delta} \right)^2, \tag{3.56}$$

where $\Delta G_I \equiv G_I - G$ and γ is the Grüneisen parameter.

The relative importance of mass-fluctuation scattering and strain scattering can be judged from the experimental data shown in Fig. 3.8. Here, the solid curve has been calculated from (3.52), and ω_D / ω_0 is determined using only the mass-fluctuation parameter A_m, given by (3.53). It will be seen that for several of the solid solution systems, the experimental data lie close to the theoretical curve. However, there is considerable disagreement in other systems. It is noteworthy that, where such disagreement occurs, the observed thermal conductivity is less than predicted theoretically on the basis of mass-fluctuation scattering. Then, it is reasonable, to suggest that the discrepancies can be attributed to a significant strain-scattering effect [3.36].

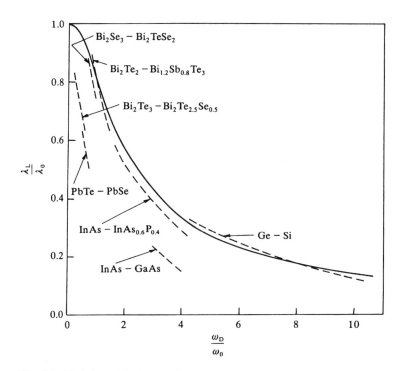

Fig. 3.8 Variation of lattice conductivity for a solid solution. The solid curve is based on (3.52), and ω_D / ω_0 is calculated only for mass-fluctuation scattering. The broken lines are the experimental data

3.4.3 Boundary Scattering of Phonons

It has been known for more than 60 years that the effect of crystal boundaries on lattice thermal conductivity occurs at very low temperatures [3.37]. However one would hardly expect it to be observed at ordinary temperatures, even for polycrystalline samples of fine grain size, because the mean free paths of phonons due to other scattering processes amount to no more than about 10^{-9} to 10^{-10} m. Therefore, it was, something of a surprise to discover that boundary scattering might be important in thermoelectric materials. The possibility was first pointed out by Goldsmid and Penn [3.38] and subsequently confirmed by experiment [3.39].

In a pure crystal at high temperatures, there are comparable contributions to the conduction of heat from all of the phonon frequencies. Because the relaxation time for umklapp scattering varies as ω^{-2}, it is clear that low-frequency phonons have a much greater free-path length than the average value and, therefore, are more sensitive to boundary scattering. Moreover, if the substance is a solid solution, alloy scattering reduces the thermal conductivity primarily by scattering the high-frequency phonons. In Fig. 3.9, the total area of the diagram represents the thermal conductivity of a very large pure crystal. The single-hatched area shows schematically the reduction of thermal conductivity due to point-defect scattering when a solid solution is formed. This means that most of the heat is now transported by the low-frequency phonons which, as we have shown, can be more easily scattered by the crystal boundaries. When the crystal size is not too large, the reduction of thermal conductivity due to boundary scattering is represented by the double-hatched area in Fig. 3.9. It is obvious that the combined effect of phonon–phonon and alloy scattering makes boundary scattering a significant effect.

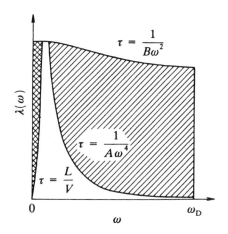

Fig. 3.9. Schematic plot of thermal conductivity against phonon frequency. The single-hatched area represents the reduction in thermal conductivity due to alloy scattering in a solid solution. The double-hatched area indicates the reduction due to boundary scattering in small crystals

Savvides and Goldsmid [3.40] demonstrated the phenomenon of boundary scattering at temperatures up to 300 K in single crystals of silicon 20 to 100 μm thick that had been neutron irradiated so as to produce numerous point defects. It was also shown that the observations could be explained by the theory of thermal conductivity in solid solutions (see Section 3.4.2) with the additional feature of an upper limit on the phonon free path length set by twice the thickness of the crystal (the so-called Casimir length).

The experiments of Savvides and Goldsmid [3.40] on fine-grained polycrystalline silicon–germanium were more relevant to practical applications because similar materials can be used in thermoelectric generators. It was found, for example, that the lattice conductivity of undoped $Si_{70}Ge_{30}$ at 300 K falls from 8.2 W m^{-1} K^{-1} to 4.3 W m^{-1} K^{-1} when the mean grain size is reduced to 2 :m. Note that the mobility of charge carriers at such a grain size is hardly reduced from that of a large crystal. On the other hand, Slack and Hussain [3.41] discussed the changes in the thermal and electrical conductivities of Si–Ge alloys that result from boundary scattering due to small grain sized specimens. Their conclusions were that a small increase in z due to the associated reduction in λ_L is obtained as the grain size is reduced to 2 μm; however, further reducing the grain sizes below this value drastically reduces z due to the reduction in σ.

It may be more difficult to produce a worthwhile effect in refrigerator materials because the thermal conductivity is already an order of magnitude smaller than that of silicon–germanium alloys. It is probable that a grain size well below 1 μm would be needed, and there might then be a significant reduction of carrier mobility, as well as thermal conductivity. There have been reports of an improvement in z, as will be discussed in Chap. 8, of $Bi_{0.5}Sb_{1.5}Te_3$ through boundary scattering of phonons in thin films [3.42], but a demonstration of the effect in bulk samples of bismuth telluride alloys has yet to be reported.

3.5 Anisotropic Thermoelements

We have seen that refrigerators based on the Ettingshausen effect have certain practical advantages over those that use the Peltier effect. In particular, a single thermomagnetic element can be designed, in principle, to operate from any source of emf, whereas a single thermocouple can be used only with a rather small specific voltage. There is also the possibility of obtaining the improved performance of a cascade merely by employing a thermomagnetic element of suitable profile. Korolyuk et al. [3.43] showed that one can achieve these same advantages, without the need for a magnetic field, whenever the thermoelectric properties are anisotropic.

Suppose, for example, that the Seebeck coefficient in a specific crystal has different values, α_ψ and $\alpha_\perp$, in the mutually perpendicular principal crystallographic directions $0X_0$ and $0Y_0$. Then, let a temperature gradient dT/dx be applied in a direction OX that makes an angle ϕ with the OX_0 axis and $(\pi/2-\phi)$ with the OY_0 axis. Then, an electric field $\mathscr{E}_y$ will appear in the direction OY at right angles to OX, and $\mathscr{E}_y$ is related to dT/dx through the expression

$$\mathscr{E} = (\alpha_\| - \alpha_\perp)\sin\phi\cos\phi\frac{dT}{dx}. \tag{3.57}$$

The ratio of the electric field to the temperature gradient is largest when $\phi = \pi/4$, whereupon

$$\mathscr{E} = \frac{1}{2}(\alpha_\| - \alpha_\perp)\frac{dT}{dx}. \tag{3.58}$$

This equation shows that, for a sample that has a much greater length in the direction OY than in the direction OX, a large thermoelectric emf can be produced from a relatively small temperature difference. In the same way, one can use an anisotropic thermoelement to produce transverse Peltier heat flow from a longitudinal electric current. Note that in these considerations, disturbances in the flow pattern near the end contacts have been ignored.

Korolyuk and his colleagues showed that anisotropic thermoelectric effects may be demonstrated conveniently using single crystals of cadmium antimonide, although the conversion efficiency is small for this and other single-phase samples. However, Babin et al. [3.44] pointed out that a synthetic two-phase material could have better anisotropic thermoelectric properties than any single crystal. They proposed using a composite material made from successive layers of two different substances, as shown by the plain and hatched regions in Fig. 3.10. Here, the direction of current flow OX is inclined at an angle ϕ to the axis of circular symmetry OX_0. The parameters of the two materials will be labeled with the subscripts 1 and 2. The layers will be assumed to be thin compared with the sample dimensions; the thickness ratio (layer 2 to layer 1) is denoted by y.

Babin and his co-workers determined the figure of merit for an anisotropic two-phase thermoelement. One of the conclusions was that the two components should have very different properties, as would be the case for a semiconductor combined with a metal. In particular, the product of thermal conductivity and electrical resistivity should be much larger for one of the components than for the other. Then, there is an optimum value for the angle ϕ that is given by

$$\tan\phi \cong \frac{y^{1/2}}{y+1}\left[\frac{\rho_1\lambda_2}{\rho_2\lambda_1}(1+Z_{12}T)\right]^4, \tag{3.59}$$

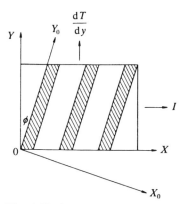

Fig. 3.10. Orientation of a synthetic anisotropic thermoelement

where Z_{12} is the figure of merit for an ordinary thermocouple composed of the elements 1 and 2 and is given by

$$Z_{12} = \frac{(\alpha_1 - \alpha_2)^2}{(\lambda_1 + y\lambda_2)(\rho_1 + \rho_2/y)}, \quad (3.60)$$

if the branches are of the same length and have areas in the ratio y. The optimum value of y is $(\rho_2\lambda_1/\rho_1\lambda_2)^{1/2}$. The figure of merit of the composite anisotropic thermoelement is given by

$$Z = \frac{(\alpha_1 - \alpha_2)^2}{\left\{(\lambda_1\rho_1)^{1/2} + \left[\lambda_2\rho_2(1+Z_{12}T)\right]^{1/2}\right\}^2}. \quad (3.61)$$

In (3.61), Z of the anisotropic thermoelement is always less than that of an ordinary thermocouple made from the same components, although the difference resides only in $Z_{12}T$ in the denominator. Samoilovich and Slipchenko [3.45] suggest that, for a particular temperature dependence of the electrical resistivity, Z for the anisotropic arrangement may be higher. However, there may be an undesirable increase in the thermal conductivity in the two-phase system because of circulating currents.

It is clear that at least one of the materials used in a composite anisotropic thermoelement should have a high figure of merit and should be selected along the lines already discussed. Thus, it is not surprising that the most convincing demonstration of the effectiveness of the artificial anisotropy concept involved the semiconductor $Bi_{0.5}Sb_{1.5}Te_3$ which, as we shall see later, is a particularly good positive thermoelectric material. Gudkin et al [3.46] made up a multilayer element from $Bi_{0.5}Sb_{1.5}Te_3$ and bismuth which, as a thermocouple, gives a figure of merit Z_{12} at 300 K equal to 0.85×10^{-3} K^{-1}. The anisotropic element, which was long enough for the contacts to have little influence on heat flow at the central region, had an optimum angle of orientation ϕ equal to 60° and a calculated figure of merit Z equal to 0.8×10^{-3} K^{-1}. The observed maximum temperature difference due to the Peltier effect was 23 K, with the hot face at 300 K, when the sample was a rectangular block. A simple cascade, consisting of a sample of trapezoidal shape, with a ratio of 10:1 between the arms of the hot and cold faces, gave a maximum temperature difference of 35 K.

3.6 Thermoelectric Cooling at Very Low Temperatures

The problem of thermoelectric cooling in the liquid helium temperature range has been considered by MacDonald et al. [3.47, 3.48] and by Blatt [3.49, 3.50], who directed attention toward metallic or semimetallic thermoelements and emphasized that only one thermoelectric material is needed, because each couple can be completed with a superconducting branch. MacDonald and his colleagues pointed out that dilute gold-based alloys have relatively large low-temperature Seebeck coefficients (for metals) of the order of 10 μV K^{-1}, and Blatt drew attention to the much larger values of α observed for bismuth in high magnetic fields, though the benefits are nullified by the large magnetoresistance effect. It has been suggested that the phonon-drag effect might lead to a high thermoelectric figure of merit at low temperatures. However, Keyes [3.51] showed that there is a rather low limit to the value of z that can be achieved using phonon drag.

Let us assume that the ordinary electronic contribution to α is small enough to be neglected compared with the phonon contribution. Then, if the Seebeck coefficient is expressed in terms of the phonon relaxation time τ_d, using (2.140),

$$z = \frac{N_d k_B x \tau_d v^2}{12 \lambda_L T}, \tag{3.62}$$

where the various symbols have been defined in Section 2.7. Here, the carrier concentration has its optimum value

$$n = \frac{N_d k_B T \mu}{3 e x \tau_d v^2}, \tag{3.63}$$

as may be found from (2.141), the electronic component of thermal conductivity is ignored.

Keyes pointed out that $N_d k_B v^2 \tau_d / 3$ is the contribution λ_d of low-energy phonons to thermal conductivity. Thus, from (3.62),

$$zT = \frac{x\lambda_d}{4\lambda_L}. \tag{3.64}$$

Because $x < 1$ and $\lambda_d < \lambda_L$, we find that $ZT < 1/4$, when phonon drag predominates. Thus, although certain simplifying assumptions have been made in obtaining (3.64), worthwhile figures of merit can never be reached by using phonon drag.

It appears rather more promising to consider instead employing ordinary electronic thermoelectric effects at temperatures that are low enough for the lattice specific heat to become negligible. Such a proposal is made practicable by the availability of helium-dilution refrigerators that can provide a continuously cooled heat sink at a temperature of the order of a few millidegrees kelvin. Of course, if thermoelectric cooling at such temperatures is to serve any useful purpose, we require it to give a further substantial depression of temperature, say, by about an order of magnitude. If this is to be achieved with a single stage, Table 1.2 shows that zT must be of the order of 20 or more over the operating range. As we shall see, this does not appear impossible.

The question of selecting of suitable materials has been discussed by Goldsmid and Gray [3.52]. They pointed out that, because the Wiedemann–Franz–Lorenz law still sets an approximate lower bound to the product $\rho\lambda$, even at low temperatures, the required high values of zT can be reached only if the Seebeck coefficient is of the order of 500 μV K^{-1}. Therefore, it is permissible to use classical statistics in the calculations, so that (3.6) is valid here. Of course, the parameter β, which is proportional to $\mu (m^*)^{3/2} T^{5/2}/\lambda_L$ also still serves the same role as that at ordinary temperatures.

If it is supposed that the temperature T is much less than θ_D, the specific heat will vary as T^3 and so, too, will the lattice thermal conductivity, if static defects or crystal boundaries are scatterers of phonons. When the charge carriers are scattered by static defects, we might expect the mobility μ to vary as $T^{-1/2}$. At low enough temperatures, then, β might eventually become proportional to T^{-1} and, if the Fermi energy could be optimized, zT should rise continuously as the temperature falls.

When $T << \theta_D$, the lattice specific heat per unit volume, is given by

$$c_V = \frac{12\pi^4 N k_B}{5}\left(\frac{T}{\theta_D}\right)^3, \tag{3.65}$$

where N is the number of atoms per unit volume. Thus, the lattice conductivity is

$$\lambda_L = \frac{4\pi^4 N k_B v l_t}{5}\left(\frac{T}{\theta_D}\right)^3, \tag{3.66}$$

3.6 Thermoelectric Cooling at Very Low Temperatures

where l_t is the mean free path of phonons. The mobility of the charge carriers is

$$\mu = \frac{4el_e}{3\left(2\pi m^* k_B T\right)^{1/2}}, \qquad (3.67)$$

where l_e is their mean free path. Substituting the values of λ_L and μ given by (3.66) and (3.67) in (3.2), we find that

$$\beta = \frac{20 k_B^2 m^* \theta_D^3 l_e}{3\pi^2 h^3 N v T l_t}. \qquad (3.68)$$

If we use the Debye approximation for the velocity of sound,

$$v \cong \left(\frac{4\pi}{3}\right)^{1/3} \frac{\theta_D}{h N^{1/3}}, \qquad (3.69)$$

we obtain

$$\beta \cong 0.13 \frac{k_B m^* \theta_D^2 l_e}{h^2 N^{3/2} T l_t}. \qquad (3.70)$$

If it is supposed that m^* is approximately equal to the free electron mass and that $l_e \cong l_t$, which could be the case if, say, both charge carriers and phonons were scattered at crystal boundaries, we find that

$$\beta \cong 4 \times 10^{-7} \frac{\theta_D^2}{T}. \qquad (3.71)$$

Clearly, θ_D should be as high as possible, but it is unlikely to exceed 1000 K. Thus, β is not expected to exceed 0.3 at a temperature of 1 K but could reach a value of 400 at 1 mK. In fact, for the optimum carrier concentration, a value of zT equal to about 20 would result from $\beta \cong 300$, so that thermoelectric refrigeration in the millidegree kelvin region begins to appear possible.

The major difficulty would lie in achieving the optimum carrier concentration. Using the value of β given by (3.68), the optimum reduced Fermi energy at 1 mK turns out to be about −5 at a corresponding carrier concentration of only 10^{15} m^{-3}. Generally, at very low temperatures, the electrons and holes would remain tightly bound to the impurity sites for small donor or acceptor concentrations. At higher concentrations, the impurity levels would broaden into bands that overlap the conduction or valence bands and lead to metallic conduction. It was suggested by

Goldsmid and Gray that one might be able to use injected carriers or, perhaps, to modulate the carrier concentration using a high magnetic field.

Another problem would undoubtedly be the very low value of the cooling power because the very high electrical resistivity would severely limit the permissible current density for any realistic sample length. On the other hand, thermal capacities at low temperatures become small, so that slow rates of heat extraction might be tolerable. Thus, thermoelectric cooling at very low temperatures cannot be discounted, though the ideas just presented remain to be demonstrated experimentally.

4. Measurement and Characterization

In this chapter, we outline the principles on which the measurement of the transport properties of thermoelectric materials are based. Then, we describe some of the practical techniques that have been used. The description is not meant to be exhaustive, and undoubtedly there are other pieces of apparatus that are at least as accurate and convenient as those that have been selected as illustrative examples. However, as in the other chapters in Part I of this text, we intend to emphasize principles rather than current practice.

4.1 Electrical Conductivity

The performance of a thermoelectric refrigerator can be expressed in terms of the figure of merit Z. Clearly, this quantity is of utmost importance. The direct determination of Z will be described in Sect. 4.4, but it is often necessary to make independent measurements of the electrical resistivity ρ, the Seebeck coefficient α, and the thermal conductivity λ. Such measurements are required, for example, in optimizing the properties of any particular semiconductor and in selecting new materials. Measurements of ρ or α are also used in routine testing of production thermoelectric material.

Although the measurement of ρ of conducting materials has been studied for a long time, there are special problems that arise for thermoelectric materials because the Peltier effect sets up temperature gradients which, in turn, generate electric fields through the Seebeck effect. Thermoelectric voltages are superimposed on the potential differences that arise from electrical resistance.

If it were not for the thermoelectric effects, ρ could be found from the electrical resistance R of a block of length L and uniform cross-sectional area A, using the relationship, $\rho = AR/L$. This, of course, is based on the assumption that there is no contribution to resistance from electrical contacts. Typical samples of thermoelectric material have resistance values less than 10 milliohms, so that the contacts must be extremely good if their effect is to be ignored completely. Therefore, it is often advisable to use a four-contact technique, as shown in Fig. 4.1, when determining ρ. The current is introduced through large-area soldered contacts at either end of the

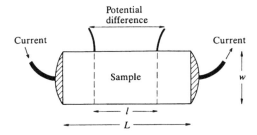

Fig. 4.1. Sample for electrical resistance measurement with four contacts

sample. The potential difference is determined across probes that are inset from the ends. The probes may be held in place by pressure, by welding, or by insertion in small holes drilled in the sample. It is usually assumed that the probes have a negligible influence on the potential distribution in the sample and that they take up the potentials corresponding to their centers, whatever their thickness.

Certain practical considerations must be taken into account [4.1]. Unless one is certain that the end faces of the sample are at uniform potential, the potential probes should be inset by at least the width w of the sample. Referring to Fig. 4.1, this means that $(L - l) > 2w$. On the other hand, in large-grained polycrystalline samples of anisotropic materials, potential probes that are too far from the ends yield unreliable results.

If the potential probes are not at the same temperature, there will be a Seebeck voltage to be taken into account. This possibility has long been recognized, and it has been customary to determine the average potential difference when the same current is passed in opposite directions. This certainly eliminates the effect of thermal asymmetry in the experimental arrangement, but it does not remove the problems originating from the Peltier effect. This difficulty was encountered and recognized by Putley [4.2] and was turned to advantage by Harman [4.3], as we shall see later.

There are two ways of avoiding errors from temperature gradients in samples. One method, which is usually impractical, is to make the probes from a substance that has precisely the same Seebeck coefficient as the sample being measured. The more usual method takes advantage of the fact that a temperature gradient, arising from the Peltier effect, takes some time to develop. Thus, it was found that either an ac current source [4.4] or a chopped dc current source [4.5] can be used to measure electrical resistance without interference from the Seebeck voltage.

4.2 Seebeck Coefficient

In principle, α is one of the simplest quantities to measure because it is defined as the ratio of an open-circuit potential difference to a temperature difference. Nevertheless, it is only too easy to introduce substantial errors in determinating it. Two possible arrangements for the measurement are shown in Fig. 4.2. the temperature difference is determined by using copper–constantan thermocouples, and the copper branches are also used to obtain the potential difference. The reference junctions of the thermocouples may be fixed to a heat sink whose temperature can be controlled. The reference temperature is measured conveniently by using a thermistor mounted on the heat sink.

If the copper output leads are connected to a programmable multiplexer, then the potential differences V_{AB}, V_{CD}, and V_{AC} can be read sequentially by a single digital voltmeter. It is good practice to vary the thermal gradient and take α as $dV_{AC}/d(\Delta T)$, rather than using a single $V_{AC}/\Delta T$ datum, to partially compensate for differences in thermocouples. It should be noted that these arrangements yield α with respect to copper, so that the absolute α of this metal (which amounts to about 2 μV K^{-1} at room temperature) must be added to obtain the absolute α of the sample.

Both approaches of Fig. 4.2 (a) and (b) have possible sources of error. In the method illustrated by Fig. 4.2(a), heat may be conducted along the wires to the sample, changing the temperature at the contact regions. Then, there may be a temperature gradient within the junction of the thermocouple wires, so that the measured temperature is different from that at the position where the electric potential is taken up. To minimize the errors from this cause, the thermocouple wires should be of very small diameter and, if necessary, a replacement for copper, with its high λ, should be found.

This problem becomes more severe as the λ of the sample becomes smaller.

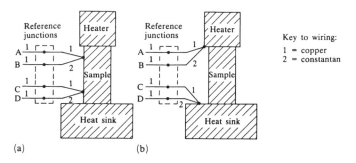

Fig. 4.2 Arrangements for measuring the Seebeck coefficient. In (**a**) the thermocouples are attached directly to the sample, whereas in (**b**) the thermocouples are attached to the source and sink

For thermoelectric materials, it is sometimes preferable to adopt the technique shown in Fig. 4.2(b). Here the thermocouples are attached to the heat source and sink, usually of copper, and λ is much higher than that of the sample. If the blocks can be soldered to the sample, there is no doubt that the true differences of potential and temperature are measured. The arrangement also works well when the source and sink are pressed against the bare faces of the sample because the entire temperature drop occurs within the thermoelectric material, despite a possible increase in contact resistance. The only undesirable feature is that the temperature gradient is highest near the interfaces, so that the regions near the ends of the sample make the greatest contribution to the observed α in nonuniform material. We note that such pressure contacts should on no account be used if the ends of the sample have been metallized. In this case, a significant portion of the temperature difference may occur outside the thermoelectric substance.

Sometimes, it is desirable to measure α in a relatively small volume of material, so that one can study the effects of inhomogeneity. An apparatus suitable for local Seebeck measurements [4.6] is shown in Fig. 4.3. A hot probe whose tip is a thermocouple junction contacts a sample that is otherwise isothermal. A second thermocouple is bonded to the sample far from the probe, and the sample is scanned beneath the probe to provide the desired spatial resolution. By measuring the voltage differences, U_1 and U_2, between the corresponding thermocouple branches, one can obtain α in the vicinity of the probe without reference to the unknown temperature difference $T_1 - T_0$. The local value of α is found from

$$\alpha = \frac{U_1}{U_2 - U_1} \alpha_{Cu/CuNi} + \alpha_{Cu}. \qquad (4.1)$$

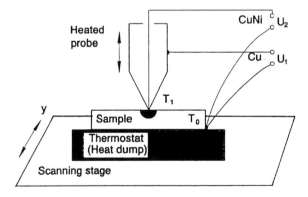

Fig. 4.3. Illustration of an apparatus used to measure local variations of the Seebeck coefficient [4.6]

The assumptions for the validity of this expression are the same as those for a bulk Seebeck measurement in which a bare sample is simply pressed between a metallic source and sink. The sample should have a much lower λ than the metallic probe, so that the temperature drop at the interface occurs predominantly in the former. A good thermoelectric material will satisfy this condition.

4.3 Thermal Conductivity

It is more difficult to measure λ than σ with accuracy because thermal insulation is never as good as electrical insulation. This means that there are invariably problems associated with heat losses. Then, the aim is to make these losses as small as possible. This is sometimes difficult when working with thermoelectric materials because they usually have low values for λ.

As we shall show in the next section, the λ of a good thermoelectric material can be measured indirectly by determining z. It is usually necessary, however, to directly measure λ in the search for new materials and during the optimization of those that already exist. The technique most frequently used to determine λ in bulk thermoelectric materials is the longitudinal steady-state method, analogous to a potentiometric measurement of ρ. Using this technique, λ can be readily measured to very low temperatures. The basic technique is illustrated schematically in Fig. 4.4. One end of the sample that has a uniform cross-sectional

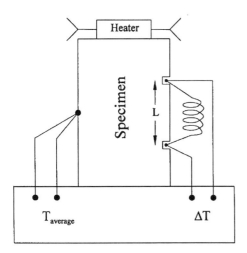

Fig. 4.4 Experimental arrangement for measuring λ by the steady-state technique

area A makes good thermal contact with the cold finger of the cryostat (via Apiezon grease or silver-epoxy, for example), and a small heater (a metal-film or a metal-chip resistor) is similarly attached to the free end of the specimen. Electrical power Q dissipated in the heater provides the heat flow through the sample. A different thermocouple (or a pair of thermometers or calibrated thermocouples) separated by a distance L measures the temperature difference ΔT along the specimen. In addition, a calibrated thermocouple (or thermometer) is also attached to accurately measure the temperature of the specimen. Then, the measured λ is calculated from

$$\lambda = \frac{QL}{A\Delta T}. \qquad (4.2)$$

If ΔT is not too large, the value of λ obtained from (4.2) will be that corresponding to the mean temperature between the differential thermocouple, even if λ is a rapidly varying function of temperature. For example, assuming $\lambda = KT^3$ (where K is a constant) even for a huge temperature difference of $\Delta T/T = 0.2$, the difference between the measured and true conductivities is only about 1%. Typically, therefore, the difference between the true λ value and that derived from (4.2) is much less than 1%.

The steady-state technique is most effectively employed when virtually all heat generated by the heater flows through the specimen to the cold sink. As λ of thermoelectric materials is generally rather low, and as will be described in Chap. 6, certain novel materials have reached glass-like λ values, care must be taken in the choice of the wires connected to the specimen and heater so that heat loss is kept to a minimum. Typically the thermocouple and heater wires are at most 25 μm in diameter. Of course, good vacuum must be maintained to eliminate heat exchange due to convection. The heat losses due to radiation will limit the maximum temperature at which this technique can be employed. In general, thermoelectric materials have relatively small band gaps and high emissivities.

Even after taking care to minimize the losses due to radiation and conduction through the wires, losses of heat by radiation and by conduction along the lead wires remain, but these are small and well defined and can be determined by conducting experiments on specimens of known small thermal conductance. These can also be estimated by remeasuring the sample after removing it from the heat sink; the heat flow then is essentially due to radiation losses and conduction through the wires. Any thermal resistance at the end contacts is found by making measurements on samples of different lengths. Experimental errors throughout the entire temperature range are typically < 3% for specimens with $\lambda \approx 2$ W m^{-1} K^{-1}.

The choice of dimensions for the sample in any thermal conductivity measurement represents a compromise between conflicting requirements. The advantage of increasing the length is that the errors from thermal resistance at the end contacts are minimized. On the other hand, the relative effect of lateral heat losses is greater for a long rather than a short sample. A short length is recommended when λ is low, to reduce the errors due to losses and also to allow establishing thermal equilibrium more rapidly [4.7].

At higher temperatures, it becomes more difficult to account for losses by radiation, and a steady-state absolute technique is less suitable. One solution is to estimate the degradation of the heat flux between the source and sink. This can be accomplished by sandwiching the sample between standards of known thermal resistance and surrounding the stack with a good thermal insulator to minimize radiation. Bowers et al. [4.8] describe such a system used to measure InSb and InAs up to 800 °C. A difficulty with such comparative methods is the necessity for the standards to nearly match the test samples in thermal conductivity.

Because of the length of time that is usually needed to reach equilibrium in a static method, a dynamic procedure may be preferred. Indeed, much of the pioneering work on thermoelectric materials carried out by Ioffe and his co-workers used a dynamic, direct measurement of λ [4.9]. Now, it is more common to obtain high-temperature thermal conductivity by measuring the thermal diffusivity κ:

$$\kappa = \frac{\lambda}{c_V}, \tag{4.3}$$

where c_V is the specific heat per unit volume.

The laser flash method is by far the most favored for measuring the κ of bulk samples at temperatures well above ambient. A typical apparatus is shown in Fig. 4.5. The sample, typically a wafer, whose diameter is much greater than the thickness (d), is irradiated on one face with pulses of laser light not more than a millisecond long. The temperature of the opposite face of the sample is monitored, perhaps with an infrared sensor as shown in Fig. 4.5. In the complete absence of heat loss from the sample, the temperature would rise monotonically to a limiting value. In a real situation, the measured temperature will peak and then return to ambient temperature.

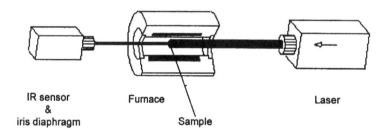

Fig. 4.5 Schematic diagram of a laser flash apparatus used for κ measurements. A wide range of κ values can be measured with such equipment

The time required to reach one-half of the peak temperature at the rear face of the specimen, $t_{1/2}$, can be used to determine κ according to

$$\kappa = \frac{1.37 d^2}{\pi^2 t_{1/2}}. \tag{4.4}$$

To approximate ideal one-dimensional heat flow, the laser spot should be uniform and have a greater area than the spot size of the temperature measurement. If a thermocouple is used to measure the temperature, then it should have a low thermal conductance or it will contribute to the reduction of the peak temperature. The sample must be thin enough that the heat pulse arrives before too much heat is lost and thick enough that the rise time is much greater than the laser pulse duration. It is advantageous that commercial laser flash systems have been developed for application to high-temperature thermal insulation materials, which have λ values comparable to good thermoelectric materials.

A dynamic technique that has been used very successfully for materials of very low thermal conductivity, such as glasses and amorphous substances, is the so-called 3ω method. It will be realized that the problem of losses through thermal radiation becomes more severe as the thermal conductivity of the sample to be measured becomes smaller. It is in this context that the 3ω method is particularly useful [4.10].

The sample consists of a slab of material with a flat surface, as in Fig. 4.6. Onto this surface is deposited a thin metal strip, with side arms for introducing current and measuring voltage. Measurements of current and voltage allow finding the power input and also, because the resistance of the strip is temperature-dependent, it enables the temperature to be determined. The strip is heated by the passage of an alternating current of angular frequency ω. Then, a temperature wave of frequency 2ω diffuses into the sample. The wave has cylindrical symmetry with the heating strip as its axis, and it is damped exponentially as it penetrates into the specimen.

The diffusion equation has been analyzed by Cahill [4.11]. The wavelength of the thermally diffusing wave, or what Cahill terms the thermal penetration depth, is a complex quantity $1/q$, which is given by

$$\frac{1}{q} = \left(\frac{\kappa}{2\omega}\right)^{1/2}. \tag{4.5}$$

Here, κ is the thermal diffusivity. The frequency must be chosen so that $1/q$ is significantly less than the thickness of the sample. Thus, the technique can be used with thin films, a key innovation as will be shown in Chap. 8, provided that the frequency is high enough. We note that in measuring potential thermoelectric materials, an insulating material (e.g., SiN) is required between the specimen and

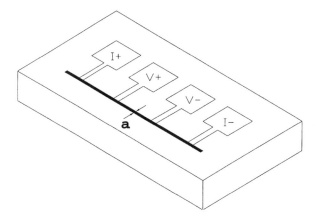

Fig. 4.6 Experimental geometry for measuring λ using the 3ω technique. The metal strip (~5 μm) simultaneously serves as a source of heat and as a thermometer

metal strip. This places an upper limit on the ω required to measure the λ of the specimen. In practice, one varies ω to determine the optimum frequency range for a particular thermoelectric material, substrate, and insulator combination.

The temperature oscillations of the heat source are determined from the third harmonic of the voltage signal; it is for this reason that the method is called the 3ω technique. Clearly, if the temperature were constant or if the resistance of the heater were independent of temperature, the third harmonic would be absent. Reference should be made to Cahill's publications for the derivation of λ from the observed quantities [4.10, 4.11].

4.4 Figure of Merit

It has already been mentioned that one can practically use the thermal emf which arises from the Peltier and Seebeck effects when a direct current is passed through a thermoelectric material. Comparison of this emf with the potential difference associated with the electrical resistance enables one to determine z directly [4.3]. This, of course, provides an alternative way of finding λ for high z materials.

Referring to Fig. 4.7a, suppose that a direct current I is passed for sufficient time to reach a steady state. If there are no thermal losses, heat is transferred from one end to the other, through the Peltier effect, at the rate $|\alpha|IT$; α is the differential Seebeck coefficient between the sample and its input leads. Ignoring the I^2R term for now, the transport of heat is balanced by thermal conduction equal to $\lambda A \Delta T/L$ such that

4. Measurement and Characterization

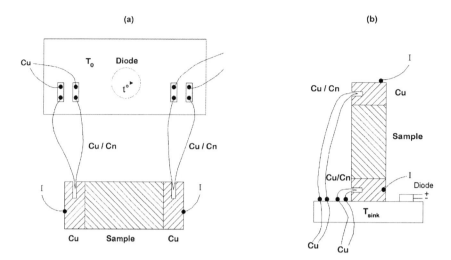

Fig. 4.7. Schematic diagrams of sample fixturing for measuring α, ρ, λ, and z. The suspended sample method (**a**) minimizes heat loss. Heat sinking the sample (**b**) allows the temperature to be varied conveniently. The thermocouples may be replaced by Cu leads if one wishes to measure ρ and z

$$|\alpha| IT = \frac{\lambda A \Delta T}{L}. \tag{4.6}$$

The temperature difference between the contacts leads to a thermoelectric emf, which is given by

$$|\alpha| \Delta T = \frac{\alpha^2 ILT}{\lambda A}. \tag{4.7}$$

In the absence of the temperature difference, that is, under isothermal conditions, the potential difference between the contacts would be

$$V_i = \frac{IL\rho}{A}. \tag{4.8}$$

Thus, the total potential difference under adiabatic conditions is

$$V_a = V_i + \frac{\alpha^2 ILT}{\lambda A} = V_i \left(1 + \frac{\alpha^2 T}{\rho \lambda}\right). \tag{4.9}$$

Upon replacing $\alpha^2/\rho\lambda$ by z, we see that

$$zT = \frac{V_a}{V_i} - 1. \tag{4.10}$$

This equation shows how readily z can be found.

The theory of Harman's method must be modified to account for heat losses. We consider a sample that is suspended by its leads (which may include thermocouples, as well as current-carrying wires) symmetrically placed at either end. It is assumed that the current is small enough to neglect Joule heating, although as we shall show later, this assumption can be relaxed without any effect on the result. The Peltier heat flow causes one end of the sample to rise to a temperature of $T_0+\Delta T/2$ and the other end to fall to $T_0-\Delta T/2$, where T_0 is the ambient temperature. The following heat losses occur:

1. Heat transfer by radiation from the end contacts. If it is supposed that ΔT is not too large, the rate of loss or gain at each end may be written as $\beta_c A_c \Delta T/2$. Here β_c is the rate of radiation per unit area per unit temperature difference, and A_c is the surface area of each contact. If the enclosure is not evacuated, β_c should also include losses by conduction and convection through the surrounding air. When the sample is in vacuum, β_c is proportional to T_0^3.

2. Heat transfer by conduction along the leads. This may be expressed as a rate of loss or gain of $K_1\Delta T/2$ at each end, where K_1 is the thermal conductance of each set of leads.

3. Heat transfer from the surfaces of the sample. Heat will be lost at the rate $\beta P(T-T_0)$ per unit length, where β is the rate of radiation per unit area per unit temperature difference, T is the local temperature, and P is the perimeter around the cross-section of the sample.

It will be assumed that the radiation losses are small enough, so that the isothermal surfaces in the sample are nearly planar. Thus, the problem is one-dimensional, and the rate of heat transfer Q is given by

$$Q = -\lambda A \frac{dT}{dx}, \tag{4.11}$$

and

$$-\frac{dQ}{dx} = \beta P(T-T_0) = \lambda A \frac{d^2T}{dx^2}. \tag{4.12}$$

At the ends of the sample, the rate of heat transfer Q_0 is

$$\pm Q_0 = \alpha IT - \frac{\beta_c A_c \Delta T}{2} - \frac{K_1 \Delta T}{2}. \tag{4.13}$$

The boundary conditions are $Q = \pm Q_0$, when $x = \pm L/2$, and $T = T_0$, when $x = 0$ (at the center of the sample). The solution of the differential equation (4.12) is

$$T - T_0 = \pm \frac{Q_0}{(\lambda A \beta P)^{1/2}} \frac{\exp\left[(\beta P/\lambda A)^{1/2} x\right] - \exp\left[-(\beta P/\lambda A)^{1/2} x\right]}{\exp\left[\frac{1}{2}(\beta P/\lambda A)^{1/2} L\right] + \exp\left[-\frac{1}{2}(\beta P/\lambda A)^{1/2} L\right]}. \tag{4.14}$$

By expanding the exponential terms and making the approximation that the temperature gradient is close to being uniform, we obtain

$$\pm Q_0 = \frac{\lambda A}{L} \Delta T + \frac{\beta P L}{12} \Delta T. \tag{4.15}$$

Combining (4.13) and (4.15), we obtain

$$\frac{\alpha IT}{\Delta T} = \lambda \frac{A}{L} + \frac{\beta P L}{12} + \frac{\beta_c A_c}{2} + \frac{K_1}{2}. \tag{4.16}$$

This equation shows that one of the advantages of Harman's technique, compared with other methods of determining λ, is that the loss terms are substantially reduced. The second term on the right-hand side of (4.16) has only one-quarter of the value that it would have if there were a heat source at one end and a heat sink at the other. The third and fourth terms have been reduced by a factor of 2, quite apart from a reduction in the magnitude of A_c.

Next, we discuss the effect of increasing the current to such a level that Joule heating cannot be ignored. This means that the center of the sample no longer remains at ambient temperature. The position on the sample at which $T = T_0$ is given by $x = a$, where

$$a \Delta T = \frac{I^2 \rho L/A}{\beta P + 2(\beta_c A_c + K_1)/L}, \tag{4.17}$$

which is obtained by setting the Joule heating equal to the sum of the heat losses. Then, the radiation losses are different at the two ends; the corresponding heat-balance equation is

$$\frac{\lambda A}{L}\Delta T + \frac{\beta P}{4}\Delta T(L\pm 2a) \mp \frac{I^2 \rho L}{A} = \alpha IT - \frac{(\beta_c A_c + K_1)}{2}\Delta T\left(1\pm\frac{2a}{L}\right). \quad (4.18)$$

Eliminating a from (4.18), we find that (4.16) results again. In other words, the magnitude of the loss term is unchanged when the apparatus is operated at currents that are high enough for appreciable Joule heating. Whether there is significant Joule heating, the modified form of (4.10) when there are heat losses is

$$zT = \left(\frac{V_a}{V_i}-1\right)\left(1+\frac{\beta PL^2}{12\lambda A}+\frac{\beta_c A_c L}{2\lambda A}+\frac{K_1 L}{2\lambda A}\right). \quad (4.19)$$

Of course, it is possible to use the Peltier effect to provide the temperature gradient for measuring either λ or z, even when one end of the sample is held in contact with a heat sink (Fig. 4.7b). This introduces asymmetry to the arrangement so that if, for example, there is significant Joule heating, the value of ΔT changes when the direction of the current is reversed. In the absence of heat losses, (4.6) still holds, provided that ΔT is regarded as the average temperature difference for the two current directions. Likewise, (4.10) can be used if V_a is taken as the average of the potential differences that are observed under adiabatic conditions for the two directions of current.

When there are heat losses, the problem may be complicated by the fact that the heat sink need not be at ambient temperature. Nevertheless, it turns out that the heat-loss terms are not affected by changes in the ambient temperature, provided that the average temperature difference ΔT for the two current directions is employed.

We find that

$$\frac{\alpha IT}{\Delta T} = \lambda\frac{A}{L} + \frac{\beta PL}{3} + \beta_c A_c + K_1, \quad (4.20)$$

when one end of the sample is in contact with a heat sink. We may compare this equation with (4.16), which holds for a suspended sample. It will be seen that the loss terms in (4.20) are virtually as large as they would be in a conventional measurement of λ. Nevertheless, such a method may be favored because attaching the sample to a heat sink greatly assists in establishing steady-state conditions.

Several versions of Harman's technique have been described in the literature. For example, it is possible to determine all three of the parameters that are involved in z if the adiabatic temperature difference is determined (see Fig. 4.7), as well as V_a/I and V_i/I. The resulting value of α will be reasonably accurate for samples with good z, because then the difference between V_a and V_i will be substantial.

In practice it is possible to measure both V_a and V_i without rearranging the sample because the thermoelectric voltage, $V_a - V_i$, decays rather slowly after current flow is terminated (Fig. 4.8). Thus it is possible with rapid data collection to precisely separate the thermoelectric and resistive voltages. The ratio of the former to the latter, $V_a/V_i - 1$, is equal to zT, according to (4.10). It is easier yet to separate the two voltage components if the voltage probes (thermocouple branches in Fig. 4.7) are inset [4.12]. In this case, when the current is initiated/interrupted, the immediate change in voltage $\pm V_i$ is followed by a brief voltage plateau before V_a begins to rise/decay.

We mention in passing that Harman's technique provides an excellent way to determine the resistance at the end contacts. Most methods for measuring the quality of the electrical contacts rely on the movement of a potential probe along the sample, and a discontinuity appears at the ends if there is any unwanted resistance. However, such a method is very difficult to apply when the contact resistances are as small as they are when good plating and soldering techniques have been used. If a measurement is made using a Harman-type apparatus, any resistance at the contact will appear as an apparent fall in z when the length becomes very small.

The measurement of potential differences under adiabatic and isothermal conditions can be used to determine z of a single sample and also that of a complete module. There is also a related technique in which the thermal resistance of a module is measured under the conditions of open and closed electric circuits. This method is one of those mentioned by Lisker [4.13], and Al-Obaidi and Goldsmid [4.14] and Berry [4.15] have applied it to couples.

The couple or module is held between a heat source and a sink, and the heat input needed to maintain a specific temperature difference is first measured when the electric circuit is open. Then, the thermal conductance K is the sum of the thermal

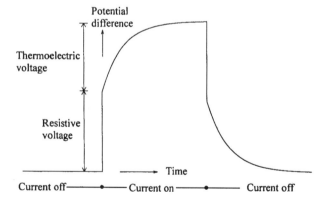

Fig. 4.8. Illustrative voltage profile for a sample in which the Peltier effect is used to establish a temperature gradient

conductances of the thermoelements. When the ends of the couple or module are short-circuited, the flow of current leads to additional heat transfer through the Peltier effect, and then K rises to K^*. The ratio of K^* to K is given by the experimental ratio of the heat input that is needed to maintain the specified temperature difference. For a temperature difference ΔT, there is a circulating current equal to $(\alpha_p - \alpha_n)\Delta T/R$, so that the Peltier heat flow is $(\alpha_p - \alpha_n)^2 T \Delta T/R$. In the open-circuit condition, the heat flow is $K\Delta T$, and we see that

$$K^* = K(1+ZT), \tag{4.21}$$

which allows finding ZT in terms of K^*/K. As with samples of thermoelectric material, it is generally necessary to account for heat losses.

4.5 Thermogalvanomagnetic Effects

Thermogalvanomagnetic effects have been studied because of the potential application of magnetothermoelectric or Ettingshausen refrigeration and also because of the information that they yield about transport mechanisms. The basic studies are usually carried out over a range of magnetic field strengths up to the highest available. A field of up to about 1 T, with a reasonable working space, can be provided by a laboratory electromagnet, but higher fields are usually provided nowadays by superconducting solenoids in liquid helium. In all cases, it is desirable that the specimen chamber be as small as possible because this assists in obtaining a high magnetic field. The basic principles have been discussed by Putley [4.2]. Here we discuss the arrangement used by Bowley et al [4.16] to study the thermomagnetic effects in Bi_2Te_3 at liquid nitrogen temperature (Fig. 4.9).

The measurements are performed with the center of the sample under transversely adiabatic conditions, by evacuating the enclosure. However, the copper blocks attached to the sample allow heat to flow in a transverse direction, which means that the potential probes and thermocouples have to be placed at some distance from the ends. It seems satisfactory if the probes that determine the longitudinal gradients are inset by at least one sample width from the ends and if the probes that determine the transverse gradients are not less than two sample widths from the ends [4.17]. The samples should be at least four times as long as they are wide. Incidentally, note that the longitudinal thermogalvanomagnetic properties, that is the magnetoresistance, the magneto-Seebeck coefficient, and the magnetothermal resistance coefficient, can be determined without using inset probes if samples of at least two different lengths are employed. It has been shown [4.18] that a coefficient C, defined for zero heat flow and current flow in the transverse direction, can be found using the relationship

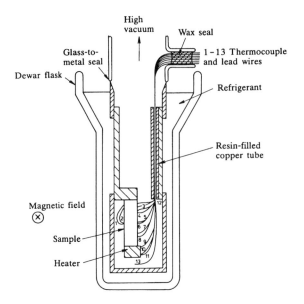

Fig. 4.9. Apparatus for determining thermogalvanomagnetic coefficients

$$C_r = \left(1 - \frac{2}{r}\right)C + \left(\frac{2}{r}\right)C^*, \tag{4.22}$$

where C_r is the value of C at a length-to-width ratio r and C^* is a constant.

Returning to the apparatus shown in Fig. 4.9, the thermocouples, designated as 1–2, 3–4, 5–6, and 7–8 are no more than 30 µm in diameter and are made from alloys, such as chromel and constantan, rather than elements, such as copper, because the latter are such good heat conductors. Magnetic alloys, such as alumel, are avoided because they might be disturbed by the application of a magnetic field. The thermocouple junctions are spark-welded to the faces of the sample; one branch of each serves as a potential probe. Wires 9 and 10 are copper current leads, and 11 and 12 are additional copper wires for determining the potential difference across the heater element, i.e., the power. Wire 11 is attached near the heater, and wire 12 near the heat sink to account for any heat generated in wires 9 and 10, which are the same length. All these copper wires are 30 µm in diameter, but wire 13 is substantially thicker. This wire is used only when a longitudinal electric current is required; it is left unconnected, for example, when magnetothermal resistance is being measured. All of the leads pass through a resin-filled copper tube which is soldered to the walls of the enclosure and forms part of the heat sink.

The thermal capacity of the heater is kept as small as possible to reduce the time to reach thermal equilibrium whenever the magnetic field, heater power, or sample current is changed. The shortness of the time interval between successive readings assists in achieving of accuracy.

The potential difference between wires 1 and 5 yields the adiabatic Nernst coefficient, whereas we would prefer to determine the isothermal coefficient. This requires taking due account of the Righi–Leduc and Seebeck effects. Likewise, the measurement of the adiabatic electrical resistivity needs to be accompanied by measurements of the Hall and Seebeck effects. It is clear that it is an advantage to determine as many of the different coefficients as possible.

Measurements of the separate coefficients, under appropriate boundary conditions, enable one to calculate the thermomagnetic figure of merit z_E. It is also possible to determine this quantity directly, as demonstrated by Guthrie and Palmer [4.19]. Their method depends on the fact that the time constant for developing any transverse temperature gradient is usually much smaller than that for a longitudinal gradient because most samples are much longer than they are wide.

Suppose that a sample of thermomagnetic material is maintained in a transverse magnetic field. The longitudinal potential difference between inset probes can be measured as a function of the time after switching on, enabling the isothermal electrical resistivity to be found. Then, the potential difference changes as a result of the temperature gradients associated with the different thermogalvanomagnetic effects, until a steady state is reached. The changes have three distinct causes:

1. The longitudinal temperature gradient due to the Peltier effect gives rise to a Seebeck voltage, as in determining z by Harman's method.
2. The transverse temperature gradient due to the Ettingshausen effect gives rise to a longitudinal gradient through the Righi–Leduc effect, and, thus, an additional longitudinal Seebeck voltage appears.
3. The transverse temperature gradient due to the Ettingshausen effect also produces a longitudinal potential difference through the Nernst effect.

Both (1) and (2) involve longitudinal temperature differences, so that the corresponding potential difference takes a relatively long time to develop. On the other hand, (3), which is related to z_E involves only a transverse potential difference, so that the potential difference associated with it develops rapidly.

The time constant for the growth of a temperature gradient is proportional to the square of the linear dimension in the same direction. For example, if the length-to-width ratio for a sample is 10:1, the longitudinal time constant is 100 times its transverse value. The procedure is to observe the first plateau in the plot of potential difference versus time; this occurs well before the appearance of a second plateau associated with the development of a longitudinal temperature gradient. Suppose that the first plateau corresponds to a voltage that is V_N below the instantaneous value V_0. The two voltages are indicated in Fig. 4.10 and might be determined experimentally from the trace of a storage oscilloscope.

The transverse temperature gradient due to the Ettingshausen effect is

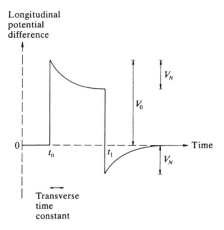

Fig. 4.10. Schematic plot of longitudinal potential difference against time in the method of Guthrie and Palmer [4.19] for determining of z_E. The current is switched on at t_0 and off at t_1; $(t_1 - t_0)$ is much less than the thermal time constant in the longitudinal direction

$$\frac{dT}{dy} = Pi_x B_z = \frac{NTi_x B_z}{\lambda}, \tag{4.23}$$

where i_x is the electric current density. The corresponding longitudinal electric field due to the Nernst effect is

$$\mathcal{E}_N = \frac{N^2 T i_x B_z^2}{\lambda}, \tag{4.24}$$

which may be compared with the electric field $\mathcal{E}_0$ associated with the isothermal electrical resistivity of the sample, given by

$$\mathcal{E}_0 = i_x \rho. \tag{4.25}$$

From (4.24) and (4.25), we find that

$$\frac{V_N}{V_0} = z^i_E T. \tag{4.26}$$

Here z^i_E is the isothermal thermomagnetic figure of merit but, as indicated previously, we prefer to use the adiabatic figure of merit z_E. We find that

$$z_E T = \frac{V_N}{V_0 - V_N}. \tag{4.27}$$

Of course, it is also possible to determine z_E from the equation

$$(T_H - T_C)_{max} = \frac{1}{2} z_E T_C^2, \tag{4.28}$$

in which $(T_H - T_C)_{max}$ is the greatest temperature difference that can be achieved for a rectangular bar in contact with a heat sink at temperature T_H. This is based on the assumption that z_E is constant over the temperature range. Because $(T_H - T_C)_{max}$ is invariably small compared with T_C, not much error is introduced.

5. Review of Established Materials and Devices

5.1 Group V_2–VI_3

5.1.1 General Properties of Bismuth Telluride

The best elemental or simple compound semiconductor for thermoelectric refrigeration at "ordinary" temperatures is bismuth telluride (Bi_2Te_3), with space group D_{3d}^5 ($R\bar{3}m$) [5.1]. Bi_2Te_3 crystals are remarkable because they can be readily cleaved in planes perpendicular to the trigonal or c axis. As one proceeds in the c direction, one encounters layers of like atoms that follow the sequence:

$$-Te^{[1]} - Bi - Te^{[2]} - Bi - Te^{[1]} -,$$

which is then repeated until a crystal boundary is reached. It has been shown that the tellurium and bismuth layers are held together by strong ionic–covalent bonds, but no bonding electrons remain to connect the adjacent $Te^{[1]}$ layers. The weak van der Waals binding between the layers accounts for the ease of cleavage [5.2].

Some of the general properties of Bi_2Te_3 are listed in Table 5.1. Note, that the observed density is significantly smaller than the value of 7.8628×10^3 kg m^{-3} that is calculated from the dimensions of the unit cell. The difference can be explained by the observation that the maximum melting composition is nonstoichiometric and contains 40.065 atomic % Bi and 59.935 atomic % Te [5.3], assuming that there are vacancies at the tellurium sites. Density measurements on samples that contain halogen impurities, which act as donors, suggest that they substitute for tellurium [5.4].

The set of elastic constants was found by Jenkins et al. [5.5] from sound-velocity measurements. Thermodynamic data up to 550°C were obtained by Bolling [5.6], and Itskevitch [5.7] determined the specific heat at low temperatures. Itskevitch's expression for specific heat contains an electronic term (depending on the first power of T) and a lattice contribution (depending on T^3). It is consistent with a Debye temperature of 155 K at 0 K.

One of the most interesting features of Bi_2Te_3 is the manner in which thermal diffusion of copper, silver, and gold takes place at relatively low temperatures. These elements act as donor impurities so that the ease with which they can move into or out of the lattice of Bi_2Te_3 is clearly of some importance. For example, it has been observed that during the manufacture of thermoelectric modules, it is possible for

Table 5.1. General properties of Bi_2Te_3 (for the composition that has the maximum melting temperature of 585°C)

Property	Symbol	Value	Temperature	Reference
Hexagonal unit Cell dimensions	a c	$(4.3835\pm0.0005) \times 10^{-10}$ m $(30.487\pm0.001) \times 10^{-10}$ m	20°C 20°C	
Density	ρ_d	$(7.8587\pm0.0002) \times 10^3$ kg m^{-3}	20°C	[5.4]
Elastic constants	c_{11} c_{66} c_{33} c_{44} c_{13} c_{14}	6.847×10^{10} N m^{-2} 2.335×10^{10} N m^{-2} 4.768×10^{10} N m^{-2} 2.738×10^{10} N m^{-2} 2.704×10^{10} N m^{-2} 1.325×10^{10} N m^{-2}	27°C 27°C 27°C 27°C 27°C 27°C	[5.5]
Specific heat (high temperatures)		$1.507 \times 10^4 + 54.4T - 0.130T^2$ J K^{-1} kg-mole^{-1}	up to 550°C	[5.6]
Specific heat (low temperatures)		$(0.84\pm0.37)T + 2.33\times10^6(T/\theta_D)^3$ J K^{-1} kg-mole^{-1}	below 2.3 K	[5.7]
Latent heat of fusion		$(1.21\pm 0.04) \times 10^8$ J kg-mole^{-1}		[5.6]
Debye temperature	θ_D	(155.5 ± 3) K	0 K	[5.7]

copper that is dissolved in molten solder to pass through nickel-plated contacts into the thermoelements; the whole process takes only a few minutes. Equally astonishing is the fact that the copper can be removed at ordinary temperatures by immersing the thermoelements in a suitable aqueous sink. The diffusion coefficients of copper in the a and c directions have been determined as a function of temperature by Carlson [5.8]. The data are shown in Fig. 5.1. The very high diffusion coefficient in the a direction is undoubtedly due to the fact that positively ionized copper atoms can readily occupy interstitial sites between the Te[1] layers. Because of the weak binding and relatively large spacing between these layers, the copper ions can also easily move from one site to another.

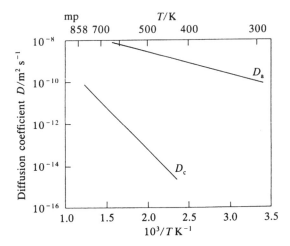

Fig. 5.1. Diffusion coefficients for copper in bismuth telluride as a function of temperature [5.8]

5.1.2 Band Structure of Bi$_2$Te$_3$

The electronic transport properties of Bi$_2$Te$_3$ are orientation–dependent but much less anisotropic than, for example, the diffusion coefficient of copper. Thus, Fig. 5.2 shows how the ratio of the electrical conductivity parallel to the cleavage planes σ_{11} to that in a perpendicular direction, σ_{33}, varies as the concentration and type of carriers are changed [5.9]. The Hall coefficient and magnetoresistance also depend on the directions of the electric and magnetic fields for both n-type and p-type crystals.

Drabble and Wolfe [5.10] showed that one could not account for the observed anisotropy of the Hall coefficient if the surfaces of constant energy in wave-vector space were centered at $k = 0$, unless the carrier relaxation time were unexpectedly anisotropic. However, it was found that measurements of the galvanomagnetic tensors in low magnetic fields could be explained, with an isotropic relaxation time, by assuming multiple energy surfaces centered at points in k space that satisfy crystal symmetry [5.11, 5.12]. The various galvanomagnetic coefficients are defined by

$$\mathcal{E}_i = \rho_{ij}i_j - \rho_{ijk}i_j B_k + \rho_{ijkl}i_j B_k B_l ,\qquad(5.1)$$

which relates the components of the electric field $\mathcal{E}$ to the components of the magnetic field B and the electric current density i. For a crystal that displays the

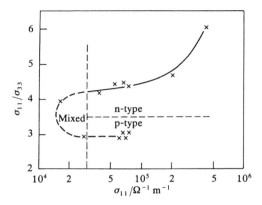

Fig. 5.2. Anisotropy of the electrical conductivity of bismuth telluride at 20°C [5.9]

symmetry of Bi_2Te_3, there are two components of the resistivity ρ_{ij}, two Hall coefficients ρ_{ijk}, and eight magnetoresistance coefficients ρ_{ijkl}.

The band structure of Bi_2Te_3 may be described with reference to the first Brillouin zone that is illustrated in Fig. 5.3. Drabble and his collaborators carried out their measurements on both n-type and p-type crystals at 77 K. They noted that in both cases the results were consistent with a six-valley model in which the band extrema lie on the reflection planes, that is, the planes containing the trigonal and bisectrix directions. The energy E of the carriers may be written as

$$E = E_0 - \frac{h^2}{2m}\left(\alpha_{11}k_1^2 + \alpha_{22}k_2^2 + \alpha_{33}k_3^2 + 2\alpha_{23}k_2k_3\right). \tag{5.2}$$

Here, α_{ij}/m is the reciprocal effective-mass tensor referred to axes in the binary, bisectrix, and trigonal directions. Experiments on galvanomagnetic effects cannot give absolute values for the components α_{ij} but only ratios between them. If all twelve of the low-field galvanomagnetic coefficients are determined, there would be redundant data allowing the assumptions to be tested for self-consistency. Thus, it was possible to establish with some certainty that the six-valley model gives a good description of both the conduction and valence band in Bi_2Te_3.

Absolute values of the components α_{ij} can be calculated once the position of the Fermi level is found. The valence band structure given in Table 5.2 represents the results obtained from Drabble's galvanomagnetic experiments supplemented by those of the magnetothermoelectric measurements carried out by Bowley et al. [5.13]. Bowley and her co-workers derived a value close to -0.5 for the scattering parameter r and thus, were able to determine the relationship between the Fermi energy and the Seebeck coefficient. The conduction band data represent the improved results of Caywood and Miller [5.14].

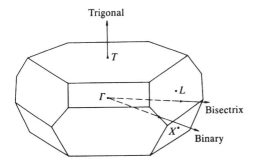

Fig. 5.3. First Brillouin zone of bismuth telluride

Table 5.2. Band structure of Bi_2Te_3 [a]

	Valence band	Conduction band
Number of valleys	6	6
Location in **k** space	On reflection planes	On reflection planes
α_{11}	19.8	26.8
α_{22}	3.26	4.12
α_{33}	4.12	3.72
α_{23}	1.0	2.4
Energy gap (eV)	$0.13 - 9.5 \times 10^{-5}(T-293)$	

[a] α_{ij}/m is the reciprocal effective-mass tensor referred to axes x and z, lying in the reflection plane, and y, lying normal to this plane.

In fact, transport measurements of the kind that have been described before cannot distinguish between a six-valley model whose extrema are inside the Brillouin zone, and a three-valley model whose extrema are at the zone boundaries. However, studies of the de-Haas–van-Alphen oscillations in magnetic susceptibility by Testardi et al. [5.15] and of the reflectance minima associated with plasma edges by Sehr and Testardi [5.16] confirmed that there are six rather than three valleys for both bands.

The data in Table 5.2 allow us to describe the appearance of the constant-energy surfaces. The small values of α_{23} indicate that the axes of all the ellipsoids are tilted slightly from the crystal axes. The similar values of α_{22} and α_{23} also show that the ellipsoids are nearly spheroidal and have a small effective mass normal to the reflection planes.

It is more difficult to carry out galvanomagnetic measurements at room temperature because carrier mobilities are then much smaller than at 77 K. Nevertheless, the ratio of the Hall coefficients, it has been observed, remain constant up to 300 K which suggests that there is little dependence of the shape of the energy bands on temperature. Some temperature dependence of the effective mass, as proposed by Ikonnikova et al. [5.17], however, is needed to explain the variation of the Seebeck coefficient with temperature for extrinsic Bi_2Te_3 [5.18]. The shape of the constant-energy surfaces certainly changes when the carrier concentration becomes very large [5.18]. Such a change is consistent with the variation in the anisotropy of the electrical conductivity with electron concentration that is shown in Fig. 5.2. Caywood and Miller [5.14] support the proposition of Mallinson et al. [5.19] that the effect is evidence for a second conduction band whose edge is 0.03 eV above that of the first band.

There are various ways of finding the value of the energy gap. One may observe the variation of the electrical conductivity and Hall coefficient with temperature in the region of mixed conduction [5.20]. Perhaps more reliable is the determination of the position of the absorption edge in the infrared region of the spectrum. Such a determination has been carried out by Austin [5.21] who took due account of phonon-assisted transitions and of degeneracy on the shape and position of the band edge. His value of 0.13 eV for the energy gap at 20°C has been confirmed by Greenaway and Harbeke [5.22]. The energy gap becomes larger as the temperature decreases; the temperature coefficient is -9.5×10^{-5} eV K^{-1} down to -155°C.

Recent band-structure calculations by Mishra et al. [5.23] and Bartkowiak and Mahan [5.24] confirm the presence of six valleys in both the valence and conduction bands of Bi_2Te_3. Both groups agree that the valence band maxima are in mirror planes somewhat close to the line $\Gamma-T$. For the conduction band, Mishra et al. find only two valleys along the line $\Gamma-T$, but they note that a small aspherical correction to their model atomic potential could lead to sixfold degeneracy. A sixfold degeneracy is found for the conduction band using a different technique, but only if the crystal structure is relaxed [5.24]. Relativistic spin-orbit corrections are essential to the band-structure calculation [5.23, 5.24]. Without these corrections, both the valence and conduction bands have a single valley at Γ. The calculated band gap is 0.08 eV [5.23] to 0.11 eV [5.24], somewhat smaller than the experimental value.

5.1.3 Transport Properties of Bismuth Telluride

The transport properties of Bi_2Te_3 are usually measured with current flow along the basal planes because then the effect of any damage due to cleavage is minimized. Unless otherwise stated, it will be implicit that the flow of electric charge or thermal energy is in this direction.

The value of the electrical conductivity σ in the extrinsic region depends on the concentration of donor or acceptor impurities. Nevertheless, all samples display the variation of conductivity with temperature that is shown in Fig. 5.4. The donor and acceptor levels lie extremely close to the conduction and valence band edges, respectively. Thus, even at the smallest impurity concentrations that can be practically achieved, the broadening of the impurity levels prevents one from observing carrier freeze-out at low temperatures. Instead, the electrical conductivity becomes temperature-independent as scattering by static defects predominates over thermal scattering.

From the liquid nitrogen temperature upward, thermal scattering is dominant and the electrical conductivity falls rapidly as the temperature rises. For the p-type and n-type samples of lowest carrier concentration, the conductivity in this region varies as $T^{-1.95}$ and $T^{-1.68}$, respectively. It has been assumed that the carrier concentration is constant and that the conductivity variation represents the dependence on temperature of the hole or electron mobility [5.11, 5.25].

At still higher temperatures, σ reaches a minimum and eventually rises exponentially because then, most of the carriers originate from thermal excitation across the energy gap. Measurements on the purest specimens indicate that σ of intrinsic Bi_2Te_3 at 20°C is equal to $1.4 \times 10^4 \, \Omega^{-1} \, m^{-1}$ [5.18].

To determine the actual value of the hole or electron mobility, the carrier concentration must be found from Hall effect measurements. In general, the Hall coefficient R_H is related to the carrier concentration n by

$$R_H = \mp \frac{A_H}{ne}, \tag{5.3}$$

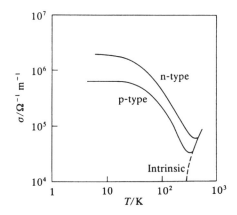

Fig 5.4. Temperature dependence of electrical conductivity in the a direction for typical speciments of bismuth telluride [5.115]

where A_H is a factor that depends on the band structure, the degree of degeneracy, and the scattering law. It is usual to measure the R_H of Bi_2Te_3 with the magnetic field in the c direction and both the current and the Hall field in the basal planes. For this orientation, the band-structure determinations of Drabble and his colleagues show that A_H is equal to 0.47 (0.33) at 77 K for degenerate samples of p-type (n-type) Bi_2Te_3. R_H in the extrinsic region does not remain completely independent of temperature. Part of this change may be attributable to a variation in A_H, as the sample becomes less degenerate. The behavior probably suggests that our description of the band structure is no more than a moderately good approximation. The values of carrier mobilities at 20°C, given in Table 5.3, were obtained by Drabble [5.11] on the assumption that the hole and electron concentrations remained unchanged from 77 K up to room temperature.

The measurement of the Seebeck coefficient is important because this quantity is involved in the thermoelectric figure of merit Z and also because it enables the Fermi energy to be deduced. Thus, by assuming that $r = -1/2$, a combination of Hall and Seebeck measurements indicates that the density-of-states effective masses for holes and electrons at 77 K are equal to $0.51m$ and $0.37m$, respectively. Just as R_H displays an unexpected temperature variation, so also does

Table 5.3. Electrical, optical, and thermal properties of bismuth telluride

Property	Symbol	Value	Temp
Carrier mobility (perpendicular to c axis):			
Electrons	μ_n	0.120 m² V⁻¹ s⁻¹	293 K
Holes	μ_p	0.051 m² V⁻¹ s⁻¹	293 K
Temperature dependence of mobility:			
Electrons		$\mu \propto T^{-1.68}$	
Holes		$\mu \propto T^{-1.95}$	
Density-of-states effective mass:			
Electrons	m_n^*	$0.37m$	77 K
	m_n^*	$0.58m$	293 K
Holes	m_p^*	$0.51m$	77 K
	m_p^*	$1.07m$	293 K
Exponent in scattering law	r	-0.5	
Refractive index		9.2	
Lattice thermal conductivity:	λ_L		
Perpendicular to c axis		1.5 W m⁻¹ K⁻¹	300 K
Parallel to c axis		0.7 W m⁻¹ K⁻¹	300 K

the Seebeck coefficient of extrinsic Bi_2Te_3 vary with temperature more rapidly than would be expected if the carrier concentration were strictly constant. Thus the density-of-states masses at 293 K have higher values of 1.07 m and 0.58 m for holes and electrons, respectively.

Figure 5.5 shows how the Seebeck coefficient varies with the electrical conductivity of Bi_2Te_3 at 20°C. Undoped bismuth telluride (which is non-stoichiometric) is p-type and has a Seebeck coefficient of about 230 µV K^{-1}. The Seebeck coefficient falls as the acceptor impurity, Pb in this case, is added, and the electrical conductivity rises. On the other hand, when iodine, a donor impurity, is added, the Seebeck coefficient rises to a maximum of about 260 µV K^{-1}, then falls to zero, changes sign, and reaches a minimum of about –270 µV K^{-1}. The changes of electrical conductivity also show that the iodine additions compensate at first for the nonstoichiometry and then cause the material to become n-type.

Goldsmid et al. [5.26] found that the Seebeck coefficient of extrinsic Bi_2Te_3 is independent of orientation, a fact that supports the assumption of an isotropic relaxation time. However, because the anisotropy of the mobility is different for holes and electrons, the Seebeck coefficient depends on the direction of the temperature gradient in the mixed and intrinsic regions.

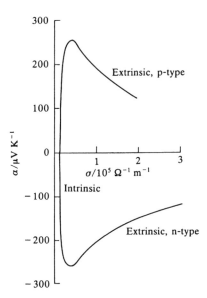

Fig. 5.5. Seebeck coefficient plotted against electrical conductivity of bismuth telluride at 20°C

The thermal conductivity λ of Bi_2Te_3 is particularly interesting in that the electronic and lattice components are of comparable magnitude [5.27]. This is apparent from the dependence of λ on the electrical conductivity σ, as shown in Fig. 5.6. At 150 K, λ tends to the value 2.7 W m^{-1} K^{-1} at zero σ, indicating, that this is the lattice component λ_L for the pure compound. However, if the electronic component is calculated, using the theory of Chap. 2, λ_L falls as σ becomes greater. It is thought that the impurities that are added to dope Bi_2Te_3 must be strong phonon scatterers; the halogens have an unexpectedly large effect [5.28].

An additional feature becomes apparent at 300 K, as indicated by Fig. 5.6 (b). Here, the bipolar thermodiffusion mechanism is so strong for mixed and intrinsic Bi_2Te_3 that λ actually rises as σ falls. In fact, for intrinsic Bi_2Te_3, λ_e/σ has the remarkably high value of 24.8 $(k_B/e)^2$ which is consistent with the prediction from the energy gap and the mobility ratio.

Measurements of λ of Bi_2Te_3 have been made from liquid helium temperature up to room temperature. The lattice component over this range, obtained by subtracting the theoretical electronic contribution, employing the Wiedemann–Franz relationship, from the total ($\lambda_L = \lambda - \lambda_e$), is shown in Fig. 5.7. The maximum at about 10 K occurs when scattering of phonons by grain boundaries takes over from phonon–phonon scattering.

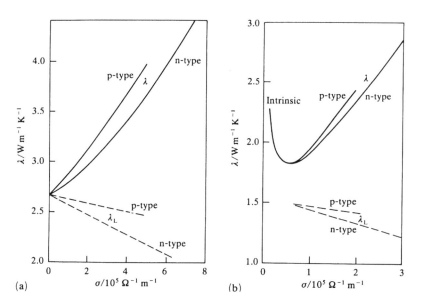

Fig. 5.6. Thermal conductivity plotted against electrical conductivity for bismuth telluride at (a) 150 K (b) 300 K

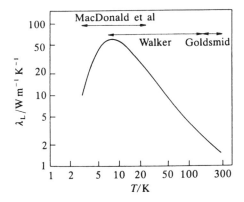

Fig. 5.7. Lattice component of λ of Bi_2Te_3 as a function of temperature. The data represent the results of MacDonald et al. [5.116], Walker [5.117], and Goldsmid [5.118]

It is rather difficult to measure λ in the c direction, but it is reliably established that the lattice component in this direction is lower by a factor of 2.1 than that in the a direction [5.28]. Because the anisotropy of the electron or hole mobility is significantly greater, it follows that $(\mu/\lambda_L)(m^*/m)^{3/2}$ is higher when transport takes place in the a rather than in the c direction. This is important when we wish to use sintered thermoelements in which it is difficult to ensure that the grains are substantially aligned in the preferred direction.

Although Bi_2Te_3 thermoelements have been superseded now by those based on solid solutions of Bi_2Te_3 with isomorphous materials, it is worthwhile to discuss the figure of merit z for the compound itself. The value of z at 300 K is greatest for a thermocouple in which both branches have an electrical resistivity of 1.0×10^{-5} Ω m; the optimum Seebeck coefficients are 185 μV K^{-1} and -205 μV K^{-1} for the p-type and n-type elements, respectively. The λ values of both branches are about 1.9 W m^{-1} K^{-1}, giving a Z for the couple of about 2.0×10^{-3} K^{-1}.

It will be seen that the Bi_2Te_3 couple exemplifies the principles that were enunciated previously. The compound has a high atomic weight, a rather low melting point, and a small Debye temperature. The highest figure of merit for both p-type and n-type material is found when the Seebeck coefficient lies close to ± 200 μV K^{-1}. There is a multivalley band structure, so that the density-of-states effective mass is substantially greater than any of the inertial masses. Although the binding has a small ionic component, it is regarded as substantially covalent, and acoustic-mode lattice scattering predominates in determining carrier mobility.

5.1.4 Antimony Telluride and Bismuth Selenide

Although neither antimony telluride (Sb_2Te_3) nor bismuth selenide (Bi_2Se_3) are particularly good thermoelectric materials, the addition of one or both to Bi_2Te_3 improves z mainly by reducing λ_L. Therefore, we shall briefly consider the properties of these two compounds, which have the same crystal structure as Bi_2Te_3.

Langhammer et al. measured the optical and electrical properties of Sb_2Te_3 [5.29]. Although the compound was prepared from elements of 99.999% purity, the samples were invariably strongly p-type and had a relatively low Seebeck coefficient of the order of 100 $\mu V\ K^{-1}$. The Seebeck coefficient was generally somewhat higher when the temperature gradient was parallel to the c axis rather than in a perpendicular direction. It was also higher when donor impurities such as iodine or excess tellurium were added. Even so, the highest value observed at room temperature was only 133 $\mu V\ K^{-1}$. σ was always higher in the a direction than in the c direction although the anisotropy ratio showed considerable variation between the samples. Lead-doped crystals displayed the largest σ of about 7 x $10^5\ \Omega^{-1}\ m^{-1}$ in the a direction and had the smallest anisotropy ratio of about 2.1. Iodine-doped or tellurium-doped crystals had σ values as low as about 3.5 x $10^5\ \Omega^{-1}\ m^{-1}$ in the a direction and an anisotropy ratio of about 2.9.

Langhammer and his colleagues found that the data from both transport measurements and reflectivity studies could be explained in terms of a six-valley valence band model. To explain the anisotropic Seebeck coefficient, an anisotropic relaxation time for the scattering of the charge carriers is needed. It is, of course, not possible to describe the conduction band from results obtained with such strongly degenerate p-type material, but the optical data on the samples of smallest carrier concentration indicate an energy gap of 0.29 eV.

A calculation of the Sb_2Te_3 band structure [5.30] finds six maxima for the valence band, located essentially at the same points in k space as the maxima of the Bi_2Te_3 valence band. The conduction band has only two valleys, but this is irrelevant because Sb-rich alloys are invariably p-type. The calculation was done on a relaxed unit cell, which had a c lattice constant 3.4% smaller than the experimental value.

For Sb_2Te_3, λ_L is difficult to determine because λ_e is always so large. Therefore, it is not surprising, that values of λ_L in the a direction at 300 K range from 1.0 W $m^{-1}\ K^{-1}$ [5.31] through 1.6 W $m^{-1}\ K^{-1}$ [5.32] to more than 2.0 W $m^{-1}\ K^{-1}$ [5.33]. All that we can really say is that the λ_L values of Sb_2Te_3 and Bi_2Te_3 are comparable.

As part of a comprehensive study of the bismuth selenotelluride system of solid solutions, Champness et al. [5.34] investigated the thermoelectric properties of Bi_2Se_3. The measurements were restricted to n-type material. This is reasonable because the alloys in this system are invariably employed as negative, rather than positive thermoelements.

Undoped Bi_2Se_3 is strongly n-type. If it were a useful thermoelectric material, it would have to be doped with an acceptor impurity to optimize the Seebeck coefficient. In fact, $(\mu/\lambda_L)(m^*/m)^{3/2}$ is no more than about one-quarter of the value for n-type Bi_2Te_3. This difference originates primarily from the electronic properties, although Birkholz and Rosi both report values for λ_L between 2.0 and 2.4 W m^{-1} K^{-1}. These values are significantly higher than the value for Bi_2Te_3.

The optical reflectivity of Bi_2Se_3 was measured by Greenaway and Harbeke [5.22] who attempted to fit a band-structure model to their data. They reported an energy gap of 0.16 eV. This is substantially less than the value of about 0.28 eV given by Austin and Sheard [5.35]. Most of the difference is attributed to the effect of degeneracy on the absorption band edge. The Seebeck coefficients of all of the samples of Greenaway and Harbeke were between −55 and −73 µV K^{-1} at room temperature, indicating that the Fermi level lies well inside the conduction band.

The optical work suggests close similarity between the band structures of Bi_2Se_3 and Bi_2Te_3. Transport measurements in a magnetic field, carried out by Caywood and Miller [5.14] on n-type bismuth selenide, show that a six-valley model can also be used for this compound, although the constant-energy surfaces are more nearly spherical than those of Bi_2Te_3.

Contrary to the experimental results, calculations indicate that the conduction band of Bi_2Se_3 has a single valley at the Γ point [5.23, 5.24]. The valence band may have either a single maximum [5.23] or six maxima [5.24], but again this is not too important because alloys with substantial Se are generally used as n-type material. The band gap computed for Bi_2Se_3 is 0.24 eV [5.23] to 0.35 eV [5.24], in agreement with the larger experimental value given before.

5.1.5 Alloys Based On Bismuth Telluride

There is complete solid solubility among the three compounds Bi_2Te_3, Sb_2Te_3, and Bi_2Se_3. A comprehensive study of the thermoelectric properties of the complete pseudoternary system has not been attempted. It turns out that the best p-type materials for thermoelectric refrigeration have compositions close to the pseudobinary system $Bi_{2-x}Sb_xTe_3$, and the best n-type materials are near the pseudobinary system $Bi_2Te_{3-y}Se_y$. Thus, these pseudobinary systems have been examined in most detail.

Undoped Bi_2Te_3 is strongly p-type. As Sb_2Te_3 is added, the number of acceptors increases still further such that in the $Bi_{2-x}Sb_xTe_3$ system, it is commonly necessary to add donor impurities to reach the optimum positive Seebeck coefficient. When x is greater than about unity it is difficult to introduce enough donor impurity for the alloy to become n-type. On the other hand, the addition of Bi_2Se_3 to Bi_2Te_3 leads to a material that is less strongly p-type. If y is large enough in the $Bi_2Te_{3-y}Se_y$ system, the alloy tends to be n-type rather than p-type. Note that factors besides composition affect the carrier concentration in the undoped material. For example, materials prepared by powder metallurgy are less strongly p-type than those produced by crystal growth from a melt.

The energy gap depends on the alloy composition. Austin and Sheard [5.35] first performed optical measurements that showed an increase in the energy gap from adding Bi_2Se_3 to Bi_2Te_3 up to the composition $Bi_2Te_{2.1}Se_{0.9}$. Further work was reported by Greenaway and Harbeke [5.22] who took due account of partial degeneracy to obtain the results shown in Fig. 5.8. It seems that adding Bi_2Se_3 is beneficial in inhibiting the onset of intrinsic conduction, a factor that is of some importance at the upper end of the temperature range at which Bi_2Te_3 alloys are used.

Although Airapetyants and Efimova [5.36] suggested that the energy gap for $Bi_{0.5}Sb_{1.5}Te_3$ is only 0.1 eV, it is not consistent with the fact that a Seebeck coefficient as high as 260 μV K^{-1} has been observed for this composition. In fact, there is no real evidence of any diminution of the energy gap in the range of x from 0 to 1.5.

Detailed studies of the thermoelectric properties of pseudobinary alloy systems were carried out by Champness et al. [5.34, 5.37], Goldsmid [5.32], Birkholz [5.33], and Rosi et al. [5.31]. Yim and Rosi [5.38] obtained slightly better thermoelectric properties than those of previous workers. All of this work was performed near or below room temperature. Results at higher temperatures were reported by Imamuddin and Dupre [5.39] who used sintered material. Better high temperature properties were found by Bulatova et al. [5.40] and by Abrikosov and Ivanova [5.41].

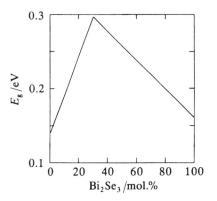

Fig. 5.8. Variation of energy gap with composition in the pseudobinary system $Bi_2Te_{3-y}Se_y$ [5.22]

Electronic measurements indicate that $\mu(m^*/m)^{3/2}$ for holes increases as Sb_2Te_3 is added to Bi_2Te_3, probably because of an increase in mobility rather than effective mass [5.36, 5.42]. The effect is not large; however, it means that there is a significant increase in σ at the optimum α. For example, at the composition $Bi_{0.5}Sb_{1.5}Te_3$, the increase amounts to nearly 20%.

The value of $\mu(m^*/m)^{3/2}$ for electrons decreases monotonically with an increase of y in the $Bi_2Te_{3-y}Se_y$ system. The results of Champness and his co-workers, however, suggest that the change is very small up to y = 0.15. We may gauge the magnitude of the effect from the variation of the room temperature Seebeck coefficient with composition when the electrical resistivity is maintained at 10^{-5} Ω m. The value of the Seebeck coefficient falls from -215 μV K^{-1} for Bi_2Te_3 and $Bi_2Te_{2.85}Se_{0.15}$ to -208 μV K^{-1} for $Bi_2Te_{2.7}Se_{0.3}$, to -200 μV K^{-1} for $Bi_2Te_{2.4}Se_{0.6}$, and to -180 μV K^{-1} for Bi_2Te_2Se. Thus alloys with more than 5 molar % of Bi_2Se_3 are likely to be attractive only if there is a fall in λ_L that more than compensates for the decrease in μ $(m^*/m)^{3/2}$ for the electrons. The high z value of 2.9 x 10^{-3} K^{-1} for n-type Bi_2Te_3 that was claimed by Fleurial et al. [5.43] is not much less than the best that has been reported for $Bi_2Te_{3-y}Se_y$, but most workers still prefer to use a solid solution rather than the pure compound.

As shown in Fig. 5.9 (a), λ_L falls with the addition of Sb_2Te_3, and reaches a minimum when molar concentration of Sb_2Te_3 amounts to about 70%. The differences between the results reported by the various workers indicate not so much any disagreement in the observed λ measurements but rather a difference in the theoretical derivation of the electronic contribution.

Different techniques for evaluating of λ_e undoubtedly account for some of the discrepancies among the results for λ_L in the $Bi_2Te_{3-y}Se_y$ system, as shown in Fig. 5.9 (b). However, the irregularities in the behavior of λ_L reported by Champness and co-workers and by Rosi and co-workers suggest that there may be real differences among the samples prepared by different laboratories. It is noteworthy, however, that the irregularities occur for molar concentrations of Bi_2Se_3 in excess of about 15%. In view of the deteriorating electronic properties beyond this composition, we are more interested in a smaller Bi_2Se_3 content. The preferred composition is usually considered to be close to $Bi_2Te_{2.7}Se_{0.3}$.

Some indication of the improvement in z that can be achieved using solid solutions, compared to that of the simple compound Bi_2Te_3, is shown in Fig. 5.10. The figures of merit z_p and z_n for the original Bi_2Te_3 samples were about 1.8 x $10^{-3}K^{-1}$ and 2.2 x 10^{-3} K^{-1}, respectively, and the values for p-type $Bi_{0.5}Sb_{1.5}Te_3$ and n-type $Bi_2Te_{2.7}Se_3$ are about 3.3 x 10^{-3} K^{-1} and 3.0 x 10^{-3} K^{-1}. The improvement in the p-type material is greater because of a larger reduction in λ_L and a rise in hole mobility for the bismuth–antimony telluride alloy.

The thermocouple Z equal to 3.15 x 10^{-3} K^{-1} is not the highest that has been reliably reported. Perhaps the best couple is that described by Yim and Rosi [5.38]. Their p-type alloy was $Bi_{0.5}Sb_{1.5}Te_{2.91}Se_{0.09}$ doped with excess tellurium, and the preferred n-type alloy was $Bi_{1.8}Sb_{0.2}Te_{2.85}Se_{0.15}$ doped with SbI_3. At 300 K, both

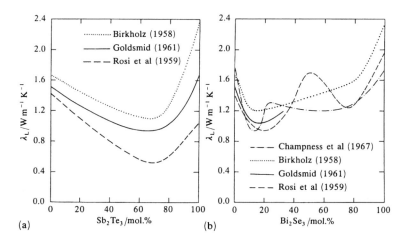

Fig. 5.9. Dependence of lattice thermal conductivity on composition in **(a)** the system $Bi_{2-x}Sb_xTe_3$ at 300 K, **(b)** the system $Bi_2Te_{3-y}Se_y$ at 300 K

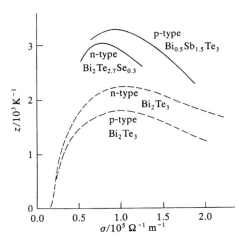

Fig. 5.10. Variation of the figure of merit with electrical conductivity for bismuth telluride and selected pseudobinary alloys at 20°C [5.119]

alloys had z values equal to 3.3×10^{-3} K^{-1}. Yim and Rosi [5.38] claimed that including selenium in the p-type alloy gave a beneficial increase in the energy gap. It is probable that there can be quite large variations in alloy compositions without much effect on z, provided that optimum doping is maintained.

We have seen that optimized alloys of Bi_2Te_3 have a zT at 300 K of about unity. Now that multistage coolers can be constructed effectively, we are also interested in what can be achieved at much lower temperatures. There are also some cooling applications, as well as thermoelectric generation from low-grade heat sources, in which the thermocouples must operate above normal ambient temperature.

First, we discuss the theoretically expected behavior of alloys in the extrinsic range of conduction. We use the fact that the optimum Seebeck coefficient is not very sensitive to changes in the parameter β. When acoustic-mode lattice scattering dominates, μ is proportional to $T^{-3/2}$, and the effective density of states in the valence or conduction is proportional to $T^{3/2}$. This means that both the optimum Seebeck coefficient and the optimum σ are more or less independent of temperature. Any variation of z must come primarily from changes in λ.

The values of λ_e are proportional to T for a given σ. The temperature variation of λ_L is more complicated in the materials in which we are interested because the phonons are scattered by both phonons and by point defects. Let us suppose, as an approximation, that λ_L can be expressed as

$$\lambda_L = \left(\frac{1}{\lambda_0} + \frac{1}{\lambda_I} \right)^{-1}, \qquad (5.4)$$

where λ_0 is the thermal conductivity of the pure (or virtual) crystal and λ_I is the thermal conductivity in the absence of umklapp scattering. Assuming that $\lambda_0 \propto T^{-1}$ and that λ_I is temperature-independent,

$$\lambda(T) = \left(\frac{T}{T_0 \lambda_0(T_0)} + \frac{1}{\lambda_I} \right)^{-1} + \frac{T \lambda_e(T_0)}{T_0}, \qquad (5.5)$$

where $\lambda_0(T_0)$ and $\lambda_e(T_0)$ are the values of these quantities at some temperature T_0 which we shall set equal to 300 K. Then, we may conclude that

$$zT = \frac{z(T_0) \lambda(T_0)}{\{T/T_0 \lambda_0(T_0) + 1/\lambda\}^{-1} + T \lambda_e(T_0)/T_0}. \qquad (5.6)$$

Approximate values of the variables in (5.6) at 300 K for the alloys used in thermoelectric refrigeration are

$\lambda(300) = 1.45$ W m^{-1}K^{-1},
$\lambda_e(300) = 0.45$ W m^{-1}K^{-1},
$\lambda_0(300) = 1.50$ W m^{-1}K^{-1},
$\lambda_l = 3.0$ W m^{-1}K^{-1}.

Above 300 K, it is necessary to account for the undesirable effect of minority carriers, as has been done, for example, by Goldsmid and Cochrane [5.44] for alloys with an energy gap of 0.15 eV.

From these considerations, the predictions are shown by the broken lines in Figs. 5.11 and 5.12. Figure 5.11 also shows the experimental data of Yim and Rosi [5.37] and of Abrikosov and Ivanova [5.41] on pseudoternary alloys with the indicated compositions. Only at the lowest temperatures are the observations substantially different from the predictions, and this is no doubt because no attempt was made to optimize the Seebeck coefficient over the entire range. For comparison, Fig. 5.11 indicates that z observed by Yim and Amith [5.45] for tin-doped Bi$_{0.88}$Sb$_{0.12}$ in a magnetic field of 0.75 T is not significantly higher than that for the optimized Bi$_2$Te$_3$ alloy at any temperature.

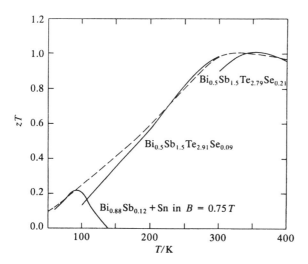

Fig. 5.11. Dimensionless thermoelectric figure of merit of p-type alloys at different temperatures. The data on the alloys of V$_2$–VI$_3$ compounds above 300 K are due to Abrikosov and Ivanova [5.41] and below 300 K to Yim and Rosi [5.38]. The low-temperature data on tin-doped bismuth–antimony alloy are due to Yim and Amith [5.45]. The broken line is the predicted behavior for optimized V$_2$–VI$_3$ alloys that give $z = 3.3 \times 10^{-3}$ K^{-1} at 300 K

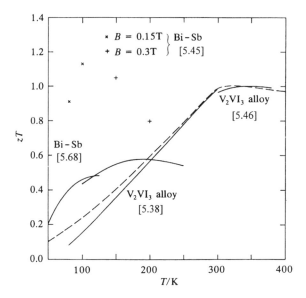

Fig. 5.12. Dimensionless thermoelectric figure of merit of n-type alloys at different temperatures. The V_2–VI_3 alloys are $Bi_2Te_{2.85}Se_{0.15}$ (above 300 K) and $Bi_{1.8}Sb_{0.2}Te_{2.85}Se_{0.15}$ (below 300 K). The broken line is the predicted behavior for optimized alloys having $z = 3.3 \times 10^{-3} K^{-1}$ at 300 K. Also shown are the data of Smith and Wolfe on bismuth–antimony alloys in a zero magnetic field ($Bi_{0.88}Sb_{0.12}$ below 125 K and $Bi_{0.95}Sb_{0.05}$ above 100 K) and of Yim and Amith on $Bi_{0.85}Sb_{0.15}$ in the magnetic fields indicated in the diagram

The experimental results of Yim and Rosi [5.38] and of Abrikosov et al. [5.46] on n-type Bi_2Te_3 alloys are compared with the experimental predictions in Fig. 5.12. Again the discrepancy at low temperatures can be attributed to nonoptimization of Yim and Rosi's alloys. Thus, there seems little doubt that optimized p- and n-type materials at temperatures below 300 K should perform at least as well as indicated by (5.6). In fact, as will be shown in the next section, n-type alloys based on Bi_2Te_3 are inferior at low temperatures to bismuth–antimony alloys. The zT values for some of these alloys, with and without a magnetic field, are also shown in Fig. 5.12.

5.1.6 Standard Thermoelectric Materials

Within the Bi_2Te_3–Sb_2Te_3–Bi_2Se_3 pseudoternary system, the optimal compositions for thermoelectric cooling are nominally $Bi_2Te_{2.7}Se_{0.3}$ and $Bi_{0.5}Sb_{1.5}Te_3$ for the n-type and p-type, respectively. When measured along the freezing direction, the thermoelectric properties of melt-grown commercial alloys (Fig 5.13) are quite similar for n- and p-type materials. To within a few percent, the room temperature properties are

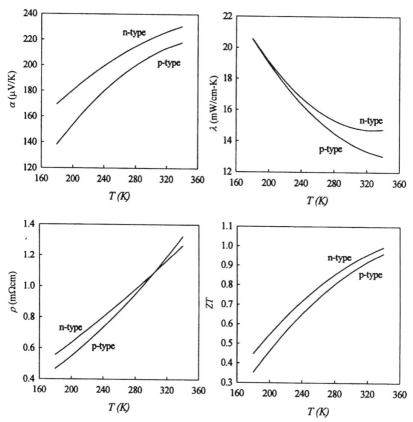

Fig. 5.13. The thermoelectric properties of typical melt-grown Bi_2Te_3-Sb_2Te_3-Bi_2Se_3 alloys. The absolute value of the n-type Seebeck coefficient is plotted

$S = \pm 210$ µV K^{-1}, $\rho = 1.0 \times 10^{-5}$ Ω m, $\lambda = 1.4$ W m^{-1} K^{-1}, and $ZT = 0.9$. Approximately two-thirds of λ is due to λ_L.

A few differences between the n-type and p-type alloys are worth noting. The temperature dependence of ρ is stronger for the p-type, for which $\rho \propto T^2$. λ_L is 10–15% higher for the n-type, but this disadvantage is offset by a higher power factor. The calculated λ_L of the n-type material also begins to increase at room temperature, probably due to bipolar electronic heat conduction, a manifestation of the transition to intrinsic conduction. A similar effect might be observed in the p-type material at slightly higher temperatures.

Perpendicular to the freezing direction, the transport properties of both n-type and p-type are less favorable. In both cases, α is nearly the same as it is along the growth axis, whereas the thermal conductivity drops to approximately 1.0–1.2 Wm^{-1} K^{-1} at room temperature. The electrical resistivity perpendicular to the freezing direction is considerably higher for both the p-type and the n-type. For the p-type material, though, the increase is not so great that z decreases drastically. On the other hand, the resistivity of the n-type material is a factor of 2–3 higher in the disfavored direction, and zT does not exceed 0.5. These differences reflect the fact that during directional freezing, the c axes of crystallites align perpendiculary to the freezing, direction. The alignment is not perfect, but the properties along the freezing direction are very nearly those of the ab plane. Perpendicular to this, the properties are an average of those for the in-plane and cross-plane directions.

5.2 Elements of Group V and Their Alloys

5.2.1 Band Structure of Bismuth and Antimony

Another materials system that requires consideration for electronic refrigeration at present includes bismuth and alloys of bismuth with antimony. Both elements have the same crystal symmetry as Bi_2Te_3 and also cleave easily along the basal planes. The Brillouin zone has features similar to those shown in Fig. 5.3 for Bi_2Te_3, but the distances between Γ and T and between Γ and L are much more nearly equal to one another and represent only a slight distortion from cubic symmetry. The structural data are usually presented in terms of a rhombohedral cell which has axes inclined to each other at 57°16, for bismuth and 57°5, for antimony.

Both bismuth and antimony are semimetals, that is, there is an energy overlap between the valence and conduction bands. The overlap is small enough for the elements to display such nonmetallic features as a change of carrier concentration when impurities are added. In fact, for bismuth, the bands overlap by only about 0.02 eV [5.47].

The conduction band of bismuth can be well represented by three ellipsoidal sets of constant-energy surfaces centered at the L-points of the Brillouin zone. Because the tilt of the ellipsoids away from the principal axes amounts to only about 5% [5.48], Abeles and Meiboom [5.49] found that they could interpret the low-field galvanomagnetic properties in terms of a structure with nontilted ellipsoids. A most noticeable feature is the very high effective mass in each valley along the bisectrix direction. The effective mass in the binary direction is rather less than in the trigonal direction and σ in the zero field is only slightly anisotropic.

Freedman and Juretschke [5.50] carried out galvanomagnetic measurements on antimony and have shown that there is also a three-valley conduction band for this element. The tilt angle is far larger than that for bismuth, and the directions of highest and lowest effective mass are interchanged.

Bismuth has a single-valley valence band with ellipsoids centered at the T-points of the Brillouin zone. Symmetry demands that the constant-energy surfaces are spheroidal about the trigonal axis. On the other hand, the valence band of antimony is of the three-valley type with constant-energy surfaces that are almost spheroidal, although tilted by about 60° from the trigonal axis.

5.2.2 Transport Properties of Bismuth and Antimony

Near room temperature, the Seebeck coefficients of both bismuth and antimony are too small for either element to be considered a competitor of the alloys based on Bi_2Te_3 in thermoelectric refrigeration. Typical values are about -70 µV K^{-1} for bismuth and 40 µV K^{-1} for antimony, although the precise values for any sample depend on its purity and crystal orientation. The addition of either the acceptor impurity, tin, or the donor impurity, tellurium, do not lead to any significant increase in the magnitude of the Seebeck coefficient, of course, because the negative energy gap implies a rather high concentration of either electrons or holes or both types of carriers. For example, Öktü and Saunders [5.51] observed that the electron and hole densities in pure antimony are nearly temperature-independent from 4.2×10^{25} m^{-3} at 273 K to 3.9×10^{25} m^{-3} at 77 K. The corresponding concentrations for pure bismuth are somewhat smaller; 2.2×10^{24} m^{-3} at 300 K and 4.6×10^{23} m^{-3} at 80 K [5.46].

The very high carrier concentration of antimony makes it unsuitable for thermomagnetic as well as thermoelectric cooling despite the comparatively high carrier mobilities that have been observed. Both types of carriers have their highest μ in the trigonal direction; values at 300 K are of the order of 0.3 m^2 V^{-1} s^{-1} for holes and 0.2 m^2 V^{-1} s^{-1} for electrons.

Bismuth is particularly interesting in that its value of $\mu(m^*/m)^{3/2}$ for electrons in the trigonal direction is greater than observed for any other material. Indeed, Gallo et al. [5.52] showed that, if the contribution of the holes to the transport processes could somehow be removed altogether, then the z at 300 K would be as high as 6×10^{-3} K^{-1}. The presence of the rather mobile holes prevents bismuth from being really useful in thermoelectric conversion, but it allows the element to be a worthwhile thermomagnetic material.

The results of the measurements by Gallo and his colleagues on bismuth are shown in Fig. 5.14 for current flow normal to and parallel to the trigonal direction. It is apparent that both the magnitude of the Seebeck coefficient and the ratio of σ to λ are higher for flow along the trigonal axis than for any other orientation. In this preferred direction, the z values at 300 K and 100 K are about 1.3×10^{-3} K^{-1} and 1.7×10^{-3} K^{-1}, respectively. The improvement in z as the temperature is lowered partly results from the rather large increase in μ, which varies approximately as T^{-2} over a wide range.

The separation of λ_e and λ_L of bismuth is difficult because there is a substantial bipolar term, in spite of the band overlap. However, the carrier mobilities are large enough for one to determine the electronic component by observing the high-field magnetothermal resistance effect [5.53]. The problem is not quite straightforward

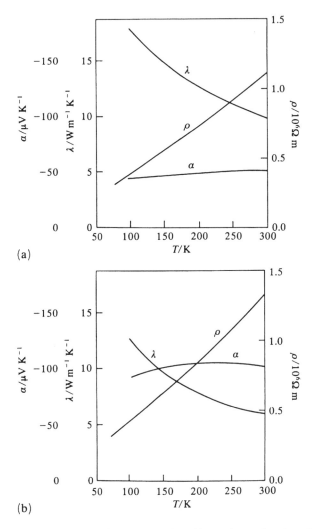

Fig. 5.14. Thermoelectric properties of bismuth as a function of temperature with current flow (a) normal to and (b) along the trigonal direction [5.52]

because λ_e does not tend to zero, as $B \to \infty$, as it does for nearly all conductors. Instead, there is a residual electronic term that is associated with transverse thermomagnetic effects [5.54]. Although it is difficult to predict the absolute value of this term precisely, one can calculate the ratio of its values for different orientations of the magnetic field. Thus, λ can be separated into its components by making measurements with the magnetic field in different directions. Figure 5.15 shows the observed temperature dependence of both λ and λ_L, where the heat flows in the binary direction. It is seen that λ is strictly proportional to T^{-1} over the range of investigation.

One of the most interesting of the transport phenomena that have been seen in bismuth is the so-called umkehr effect. This effect is manifest as a change in the value of the Seebeck coefficient in a transverse magnetic field when the direction of this field is reversed. In certain orientations, the umkehr effect can be very large, as shown by Smith et al. [5.55]. For example, in a sample of bismuth at 80 K, where heat flows along the bisectrix direction, it was found that the Seebeck coefficient was 150 µV K^{-1} when a transverse magnetic field of 1.0 T is at an angle of 60 ° to the binary direction. When this field was reversed, the Seebeck coefficient changed sign and had a value of 170 µV K^{-1}. Smith and Wolfe [5.56] pointed out that the umkehr effect should be present in any semiconductor that has nonspherical surfaces of constant energy, provided that the magnetic field does not lie along a reflection plane. However, the effect is very large in a material such as bismuth because it contains equal numbers of highly mobile electrons and holes with anisotropic effective masses. Incidentally, whenever $\alpha(B) \neq \alpha(-B)$, it is necessary to modify the second Kelvin relationship. Wolfe and Smith [5.57] have been able to confirm by experiment that $\Pi(B) = T\alpha(-B)$ so that the field direction that gives the smaller Seebeck coefficient gives the larger Peltier coefficient.

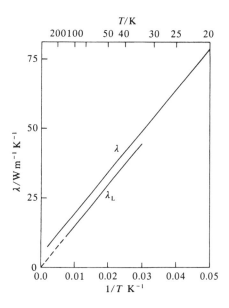

Fig. 5.15. Total and lattice thermal conductivity of bismuth in the binary direction [5.53]

The large values of the Seebeck coefficient that are found for bismuth when a magnetic field is applied suggest that z can be improved in this way. Because this z value is substantially larger for certain of the bismuth–antimony alloys, we shall defer further discussion of the magnetothermoelectric effect until we deal with these materials. Now, we shall, however, discuss the transverse thermomagnetic effects in bismuth. It is possible that pure bismuth displays the highest z value of all materials within certain ranges of temperature and magnetic field strength [5.45].

Before we review the thermomagnetic properties observed, we may attempt a theoretical prediction of z_E for bismuth in an infinitely strong magnetic field. For this purpose, we ignore the tilt of the conduction-band ellipsoids and use the electron and hole mobilities and concentration data given by Abeles and Meiboom [5.49]. In addition, we use the values for λ_L and the zero field partial Seebeck coefficients determined by Gallo et al. [5.52]. The expressions given by Tsidil'kovskii [5.58] and Abeles and Meiboom have been used to calculate the high-field values for the partial Seebeck coefficients and the magnetoconductivity, respectively. The basic parameters are given in Table 5.4 for the temperatures 80 and 300 K.

It is convenient to use the expression for the high-field Nernst coefficient N_{ji} in terms of the magnetoconductivity that has been given by Kooi [5.59]. This expression, which is applicable to an intrinsic conductor, is

$$N_{ji}B^2 = \frac{ne\left[\alpha_p(B) - \alpha_n(B)\right]}{\sigma_{ij}(B)}, \tag{5.7}$$

where $\sigma_{ij}(B)$ is the electrical conductivity in the j direction in the high magnetic field B and $N_{ji} = \mathcal{E}_j/B\nabla_i T$. The magnetic field is perpendicular to both the i and j directions. Equation (5.7) leads to an expression for the thermomagnetic figure of merit z_E,

$$z_E = \frac{n^2 e^2 \left[\alpha_p(B) - \alpha_n(B)\right]^2}{\sigma_{ij}(B)\lambda_{L,i}B^2}, \tag{5.8}$$

where $\lambda_{L,i}$ is the lattice thermal conductivity in the i direction.

The predicted values for z_E are given in Table 5.5. The preferred orientation requires the current to flow along the trigonal direction, with the magnetic field along a bisectrix direction, and the resultant Ettingshausen heat flow along a binary direction. With this orientation, $z_E T$ is equal to about 0.17 at 80 K and 0.87 at 300 K. It must be remembered that these values are predicted for an infinitely high magnetic field and that this condition is approached only when $\mu_p(1)\mu_p(3) B^2 \gg 1$. At 80 K, the requirement reduces to $B^2 \gg (0.16)^2$ T^2, which should not be too difficult to achieve, by using a SmCo$_5$ permanent magnet, for example, but at 300 K, the

Table 5.4. Basic parameters for bismuth at 80 and 300 K

Parameter	80 K	300 K
ne (C m^{-3})	7.37 x 10^{-4}	3.52 x 10^3
Electron mobilities: [a]		
$\mu_n(1)$ (m^2 V^{-1} s^{-1})	55.7	3.18
$\mu_n(2)$ (m^2 V^{-1} s^{-1})	1.40	0.08
$\mu_n(3)$ (m^2 V^{-1} s^{-1})	33.3	1.90
Hole mobilities: [a]		
$\mu_p(1) = \mu_p(2)$ (m^2 V^{-1} s^{-1})	12.4	0.77
$\mu_p(3)$ (m^2 V^{-1} s^{-1})	3.33	0.21
Partial Seebeck coefficients ($B \to \infty$):		
α_n (μV K^{-1})	−100	−125
α_p (μV K^{-1})	105	107
Lattice conductivities:		
λ_L($\perp$ trigonal axis) (W m^{-1} K^{-1})	11.0	2.9[b]
λ_L(trigonal axis) (W m^{-1} K^{-1})	7.5	2.0[b]

[a] The 1, 2, and 3 directions lie along the binary, bisectrix, and trigonal axes, respectively, in the Brillouin zone.

[b] In view of the difficulty of extracting the value of λ_L at 300 K, these values have been derived from those at 80 K, assuming that $\lambda_L \propto T^{-1}$.

requirement becomes $B^2 \gg (2.5)^2 T^2$, which is hardly a practical proposition because it could be met only using either a large superconducting magnet or a conventional high-field coil that consumes enormous amounts of power. Despite the potentially large value of $z_E T$ in bismuth, thermomagnetic cooling can only be considered competitive with thermoelectric cooling at well below room temperature.

To turn now to what has been achieved in practice, Yim and Amith [5.45] measured the value of z_E for bismuth between 70 and 300 K in a magnetic field of 0.75 T. They confirmed that the predicted orientation is the one that gives the best results. Not surprisingly, the value of $z_E T$ at 300 K is less than 0.025 because the high-field condition is so far from being met. At 80 K, $z_E T$ is about 0.24, which is somewhat larger than the predicted value.

Table 5.5. Predicted thermomagnetic figures of merit for bismuth

Orientation[a]		Figure of merit z_E for $B \to \infty$	
i	j	80 K	300 K
Bisectrix	Binary	6.3×10^{-4} K^{-1}	8.2×10^{-4} K^{-1}
Binary	Bisectrix	6.3×10^{-4} K^{-1}	8.2×10^{-4} K^{-1}
Binary	Trigonal	21×10^{-4} K^{-1}	29×10^{-4} K^{-1}
Bisectrix	Trigonal	8.4×10^{-4} K^{-1}	11×10^{-4} K^{-1}
Trigonal	Binary	12×10^{-4} K^{-1}	18×10^{-4} K^{-1}
Trigonal	Bisectrix	12×10^{-4} K^{-1}	18×10^{-4} K^{-1}

[a] i is the direction of the temperature gradient that results from an electric current in the j direction, and the magnetic field is in the third perpendicular direction.

5.2.3 Bismuth–Antimony Alloys

The electrical properties of alloys of bismuth and antimony were first studied by Smith [5.60] with polycrystalline samples of doubtful purity. A revival of interest in these materials followed the experiments of Jain [5.47] on single crystals. Jain [5.47] proposed that the alloys containing between 4 and 40 molar % of antimony are semiconductors, and have positive values for the energy gap. He claimed to observe a maximum value for the gap of about 0.014 eV at the composition $Bi_{0.88}Sb_{0.12}$. It was postulated that the position of the extremum of the valence band in the Brillouin zone changes from the T point to the L point as the concentration of antimony rises above 12%.

More precise details of the changes in the band structure have been given by Lenoir et al. [5.61] and are shown in Fig. 5.16. There is actually a positive gap at the L points, even for pure bismuth. The gap, however, falls to zero at 4% antimony and, for higher concentrations, it opens up again as the bands invert. The overlap with the valence band at the T points disappears at 7% antimony, so that above this concentration, the alloys are truly semiconducting. A positive gap exists until the antimony concentration reaches 22%, when the conduction-band minimum overlaps a valence-band maximum at the so-called H point. The widest gap occurs when the antimony concentration is about 16%. Early structural measurements on the bismuth–antimony system were carried out by Bowen and Morris Jones [5.62] and Ehret and Abramson [5.63]. The plots of the hexagonal cell parameters a and c against antimony concentration in Fig. 5.17 are due to Jain [5.47]. Noteworthy is the unusual behavior of the c parameter at antimony concentrations below 10 molar %, which may be related to the irregular changes in the energy gap.

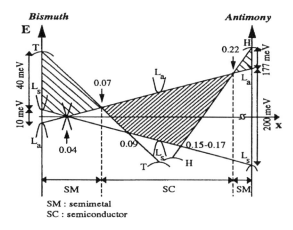

Fig. 5.16. Band structure in the bismuth–antimony system. Reprinted from [5.61] (©1996 IEEE)

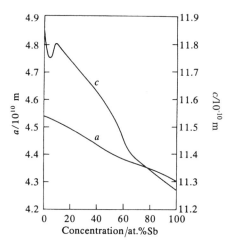

Fig. 5.17. Hexagonal cell parameters as a function of composition for bismuth–antimony alloys [5.47]

The cyclotron resonance studies by Smith [5.64] on $Bi_{0.95}Sb_{0.05}$ give an indication of the way in which the electron effective mass varies from its value for bismuth when antimony is added. The ratios between the components of the effective mass tensor do not change very greatly, but their actual values are approximately halved for the alloy. These results are consistent with a nonparabolic conduction band; the Fermi level moves downward with respect to the band edge when antimony is added.

Smith's results and their interpretation were, at least qualitatively, confirmed by Brandt et al. [5.65] for a range of alloys containing 1.7 to 4 molar % antimony. Unfortunately, the decreasing effective mass is not accompanied by an increase in carrier mobilities. For example, at 77 K, Jain [5.47] found that the average of the electron and hole mobilities is about the same for bismuth and $Bi_{0.95}Sb_{0.05}$ and that this average μ falls by a factor of about 2 at the composition $Bi_{0.93}Sb_{0.07}$.

This, of course, does not mean that the bismuth–antimony alloys are inferior to pure bismuth for electronic refrigeration because λ_L should be significantly smaller. This is clear from the measurements by Cuff et al. [5.66] on the variation of λ with transverse magnetic field strength for various bismuth–antimony alloys at 80 K, as shown in Fig. 5.18. Because the orientation of the samples is that which yields the maximum z_E, λ is not expected to approach the lattice component at a high magnetic field. There is no doubt, however, that λ_L falls with the rise in antimony content. All these alloys must have values of λ_L in the binary direction at 80 K that are much less than the value of 11 W m^{-1} K^{-1} for bismuth. The observations of Horst and Williams [5.67] for the same temperature and direction of heat flow are more valuable because the transverse magnetic field was applied in two different directions. Their work indicates a value for λ_L of about 3.7 W m^{-1} K^{-1} for $Bi_{0.95}Sb_{0.05}$ and 3.1 W m^{-1} K^{-1} for $Bi_{0.88}Sb_{0.12}$. Magnetothermal resistance measurements at higher temperatures are less useful in determining λ because of the lower μ values.

The λ_L value is smaller in the trigonal direction for both bismuth and the bismuth–antimony alloys. With heat flow in this direction, the transverse

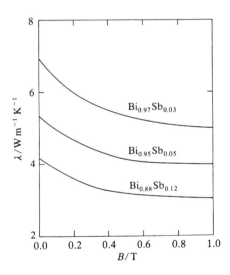

Fig. 5.18. Thermal conductivity plotted against magnetic field for various bismuth–antimony alloys at 80 K [5.66]. The temperature gradient lies in the binary direction, and the magnetic field is in the bisectrix direction

thermomagnetic effects are less significant, so that more reliance can be placed on separating the components of λ using a magnetic field. The data of Horst and Williams at 80 K indicate values of λ_L equal to about 2.1 W m^{-1} K^{-1} for $Bi_{0.95}Sb_{0.05}$, 1.7 W m^{-1} K^{-1} for $Bi_{0.92}Sb_{0.08}$, and 1.5 W m^{-1} K^{-1} for $Bi_{0.88}Sb_{0.12}$. The last of these values is three-quarters of that which one estimates from the data of Yim and Amith [5.45], which also indicate a value of 1.7 W m^{-1} K^{-1} for $Bi_{0.85}Sb_{0.15}$ in the same trigonal direction.

The z value of certain bismuth–antimony alloys is higher than that of pure bismuth because of both the reduced λ_L and the appearance of a positive, albeit small, energy gap. This gap is really too small to allow reaching a Seebeck coefficient at room temperature of much more than about 100 μV K^{-1}, but larger values are observed at lower temperatures, as Smith and Wolfe first demonstrated [5.68]. In their experiments, Smith and Wolfe found that z is higher for undoped rather than doped alloys, as evident from Figs. 5.19 and 5.20. Figure 5.19 shows the variation of the Seebeck coefficient with temperature for $Bi_{0.95}Sb_{0.05}$ (an alloy with close to a zero energy gap) and $Bi_{0.88}Sb_{0.12}$ (with one of the larger gaps). The value of -160 μV K^{-1} at 80 K for $Bi_{0.88}Sb_{0.12}$ is a considerable improvement on that for bismuth. Figure 5.20 shows the way in which the electrical resistivity varies with temperature for the same samples with the same orientation, that is, with transport in the trigonal direction. It is apparent from these figures that the lead-doped sample of $Bi_{0.88}Sb_{0.12}$ shows little difference in its Seebeck coefficient from that of the undoped samples above 60 K, but below 150 K, its electrical resistivity is significantly higher, and it is obviously an inferior thermoelectric material. Undoped $Bi_{0.88}Sb_{0.12}$ has a z of 5.2 x 10^{-3} K^{-1} at 80 K, but the value falls to 2.0 x 10^{-3} K^{-1} at 200 K and to only

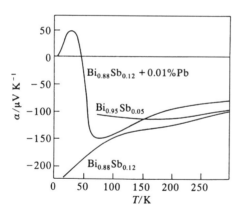

Fig. 5.19. Seebeck coefficient plotted against temperature for $Bi_{0.95}Sb_{0.05}$ and $Bi_{0.88}Sb_{0.12}$ [5.68]. The temperature gradient is in the trigonal direction

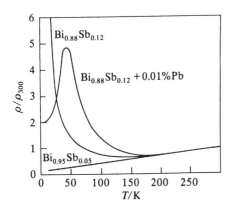

Fig. 5.20 Change of electrical resistivity with temperature for $Bi_{0.95}Sb_{0.05}$ and $Bi_{0.88}Sb_{0.12}$ [5.68]. The electrical field is parallel to the trigonal axis

1.0×10^{-3} K^{-1} at 300 K. $Bi_{0.95}Sb_{0.05}$ is slightly inferior at 80 K, where $z = 4.8 \times 10^{-3}$ K^{-1}, but superior at 300 K, where the value is 1.8×10^{-3} K^{-1}.

The z value can be substantially improved by applying a transverse magnetic field [5.69]. The highest value is obtained when the flow of heat and electric current is along the trigonal direction and the magnetic field is along a bisectrix axis. The improved properties can be seen in Fig. 5.21; z rises from 3.4×10^{-3} K^{-1} in a zero field to 7.6×10^{-3} K^{-1} in a field of 0.6 T at 160 K. The rise of α and the fall of λ more than compensate for the increase in the electrical resistivity until still higher magnetic fields are reached. The optimum magnetic field strength falls as the temperature is reduced and becomes equal to 0.25 T at 125 K and only 0.10 T at 100 K.

More recent work by Jandl and Birkholz [5.70] show some improvement for tin-doped $Bi_{0.95}Sb_{0.05}$ over undoped crystals. These workers added up to 440 ppm of tin, but found that a sample with 145 ppm has the best z. A value as high as 5×10^{-3} K^{-1} for z in the trigonal direction was observed in a magnetic field of 0.9 T at a temperature of 293 K and, even in a zero field, $z = 3 \times 10^{-3}$ K^{-1} was measured. If these values are reproducible, there must be some improvement over the best that can be achieved for n-type bismuth telluride alloys, if a moderate magnetic field can be applied. However, bismuth–antimony aligned with current flow along the trigonal direction is a much less robust material than Bi_2Te_3 in its preferred orientation. Perhaps the most significant finding by Jandl and Birkholz [5.70] is that undoped bismuth–antimony is not necessarily superior to doped material.

Note that the change of the Seebeck coefficient in a magnetic field can be extremely large. For example, from Fig. 5.19, we note that a field of 1.4 T changes the Seebeck coefficient from -130 μV K^{-1} to -260 μV K^{-1}. This is much greater than the changes that are usually observed for semiconductors. Wolfe and Smith [5.69] suggested that it might be associated with transverse thermomagnetic effects. Then, one would expect the magneto-Seebeck effect to depend on the length-to-width

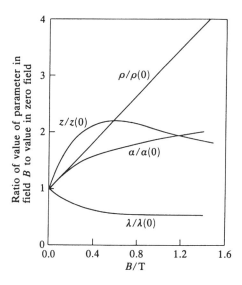

Fig. 5.21. Changes in the thermoelectric properties of $Bi_{0.88}Sb_{0.12}$ at 160 K in a magnetic field [5.69]. The transport is along the trigonal direction, and the magnetic field is along a bisectrix axis

ratio of the sample, just as the magnetoresistance effect is shape-dependent. Ertl et al. [5.71] showed that this shape dependence does occur. These results are illustrated in Fig. 5.22 for $Bi_{0.93}Sb_{0.07}$ at 80 K.

One would not expect the transverse thermomagnetic phenomena to be large enough to have much influence on the magneto-Seebeck effect of extrinsic samples. Thomas and Goldsmid [5.72] found that the magneto-Seebeck effect of $Bi_{0.95}Sb_{0.05}$ becomes smaller when there is only one type of carrier. Nevertheless, the observed change from -60 to -123 $\mu V\ K^{-1}$, when the magnetic field was increased from zero to 1.6 T, is still remarkably large and is a strong indication of a nonparabolic dispersion law for charge carriers. Note that these results were obtained when the temperature gradient was in the binary direction and the magnetic field in a bisectrix direction.

Bismuth–antimony alloys have been considered for use as p-type thermoelements. Yim and Amith [5.45] reported measurements on the alloy $Bi_{0.88}Sb_{0.12}$ doped with 300 ppm tin. The heat and electrical conduction was in the trigonal direction, and a transverse magnetic field of up to 0.75 T was applied in a bisectrix direction. In a zero magnetic field, the material displayed a p-type z of about $0.3 \times 10^{-3}\ K^{-1}$ at 90 K, but at the highest magnetic field, z rose to about $2.3 \times 10^{-3}\ K^{-1}$ at 85 K. Jandl and Birkholz [5.70] were unable to improve on this value, and it is probably close to the best that can be achieved in the bismuth–antimony system. It is also comparable with the best that can be obtained with p-type bismuth telluride alloys at low temperatures.

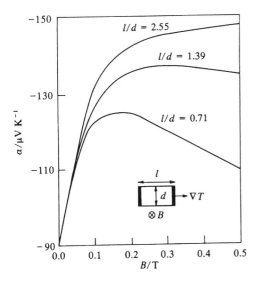

Fig. 5.22. Shape dependence of the magneto-Seebeck effect for $Bi_{0.93}Sb_{0.07}$ at 80 K [5.71]. The heat flow is along the trigonal axis, and the magnetic field is in a bisectrix direction

The need for a substantial magnetic field is a distinct disadvantage. Thus the best low temperature thermocouples use bismuth telluride alloys for p-type branches and bismuth-antimony alloys for n-type branches.

Although bismuth–antimony alloys are clearly superior to pure bismuth in thermocouples, it is not quite so evident that they are better for use in thermomagnetic elements because μ is doubly important in thermomagnetic refrigeration. Figure 5.23 shows how z_E depends on magnetic field strength. Even though the carrier mobilities are closer to those in bismuth than they would be in alloys with greater antimony concentrations, the value of z_E is below the high-field saturation value at 78 K when B is equal to 0.8 T, and well below the saturation value at 115 K and 148 K. It turns out that $Bi_{0.99}Sb_{0.01}$ is somewhat better than pure bismuth below a temperature of 130 K, if the magnetic field is restricted to, say, 0.75 T, but, as shown in Fig. 5.24, bismuth is superior at higher temperatures.

Figure 5.24 does not show the best z_E values that can be achieved. Significantly higher values have been reported by Horst and Williams [5.73]. Figure 5.25 indicates how $z_E T$ depends on temperature for a crystal of $Bi_{0.97}Sb_{0.03}$ in which the excess carrier density has been reduced to 10^{20} m^{-3}. The value of $z_E T$ is no less than 1.0 at 150 K in a magnetic field of 1.0 T. Figure 5.26 shows the strength of the magnetic field that is needed to obtain the results portrayed in Fig. 5.25 at other temperatures. It was stated that a reduction of 30% in the magnetic field, below the values given in Fig 5.26, leads to a fall of only 10% in the figure of merit.

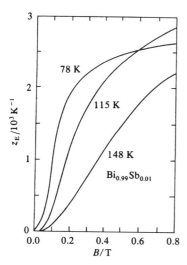

Fig. 5.23. Dependence of the thermomagnetic figure of merit on magnetic field strength for $Bi_{0.99}Sb_{0.01}$ [5.45]. The heat flow is in the binary direction, and the electric current is in the trigonal direction

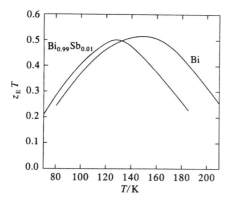

Fig. 5.24. Dimensionless thermomagnetic figure of merit of bismuth and $Bi_{0.99}Sb_{0.01}$ in a magnetic field of 0.75 T at different temperatures [5.45]

The broken curve in Fig. 5.25 shows the values for $z_E T$ calculated from measurements made by Horst and Williams [5.73] on another crystal of $Bi_{0.97}Sb_{0.03}$ that had only 10^{16} excess carriers per m^3. In this case, $z_E T$ approaches 2 at a temperature of about 100 K, but such a high value remains to be confirmed experimentally.

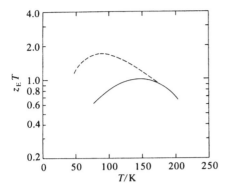

Fig. 5.25. Dimensionless thermomagnetic figure of merit for $Bi_{0.97}Sb_{0.03}$ in the preferred orientation, according to Horst and Williams [5.73]. The solid curve represents the results observed for a crystal with 10^{20} excess carriers per m^3 and the broken curve has been calculated from data for a crystal with 10^{16} excess carriers per m^3

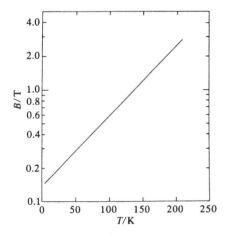

Fig. 5.26. Magnetic field that is required to achieve the values of $z_E T$ that are indicated by the solid curve in Fig. 5.25

A direct comparison between the techniques of thermoelectric and thermomagnetic refrigeration is complicated by the fact that Z of a couple is no longer approximately equal to the mean of z_p and z_n for the two branches, when the properties of the latter are very different from one another. For example, at a temperature of 100 K, the average $z_n T$ for $Bi_{0.85}Sb_{0.15}$ and $z_p T$ for a bismuth telluride alloy is about 0.65, whereas the value of ZT for the couple is only 0.45. In spite of this difficulty, it can be stated with confidence that thermomagnetic refrigeration is the superior method below about 150 K, provided that a moderate magnetic field is available, even if the sample is rectangular. When one takes account of the advantage of cascading by shape for a thermomagnetic device, the merits of the latter are even more obvious.

5.3 Materials for Thermoelectric Generators

Although the same concept of figure of merit applies for both generation and refrigeration, a number of factors are specific to generator materials. Thermodynamic considerations indicate the advantage of as high a source temperature as possible, though from the practical viewpoint, one may need to work at quite low temperatures. For example, if the intention is to use the temperature gradients in the oceans, the materials for a generator are essentially the same as needed for refrigeration. Nevertheless, much higher temperatures are needed for most generator applications.

For any given material, zT for optimum doping usually rises with temperature until it is no longer possible to prevent two-carrier conduction. Eventually, certain materials also become chemically unstable as the temperature is raised, even though it may not be close to the melting point. Thus, Bi_2Te_3 and its alloys are not suitable for use at temperatures of much more than about 400 K. Alloys based on lead telluride are inferior to bismuth telluride alloys for refrigeration, but they can be used up to higher temperatures and, with other tellurides, have been used for generation using moderately hot sources. At still more elevated temperatures, alloys of silicon with germanium have been employed successfully.

In this section, we shall discuss the properties of silicon–germanium alloys and the tellurides that have been used in thermoelectric generation. It might also be mentioned that other materials, such as iron disilicide, have been used when conversion efficiency has been of secondary importance [5.74], but we shall concentrate on those materials that display high figures of merit at their temperatures of operation. In Part II of the text, we shall discuss some of the new materials that are already comparable with tellurides and silicon–germanium alloys and may lead to superior figures of merit for both generation and refrigeration in the future.

5.3.1 Lead Telluride and Related Compounds

When Ioffe and his colleagues led the revival of interest in thermoelectricity in the middle of the twentieth century, one of the first materials that they studied was lead telluride (PbTe) [5.75]. It clearly meets the high atomic weight criterion and has a low λ_L combined with reasonably good electronic properties. Its solid solutions also were the first that demonstrated the principle that λ_L could be reduced without appreciably diminishing μ [5.76].

PbTe melts at a higher temperature than Bi_2Te_3, 923°C compared with 585°C. It also has a cubic sodium chloride crystal structure, which means that the thermoelectric properties are the same in all directions. Thus, randomly oriented polycrystalline material might be just as good as a single crystal. The band gap of 0.32 eV at 300 K allows the Seebeck coefficient to rise to a value in excess of 300 $\mu V\ K^{-1}$ after suitable doping. Both n-type and p-type samples can be produced by departures from stoichiometry [5.77], and it is also possible to introduce impurities such as the alkali metals, that act as acceptors, and the halogens, which are donors. PbTe has the rather high μ_n of 0.16 $m^2\ V^{-1}\ s^{-1}$ at 295 K, and μ_p is 0.075 $m^2\ V^{-1}\ s^{-1}$ [5.78]. The effective masses of both carriers are equal to about 0.03 m. Note that μ_p is marginally higher for lead selenide at 0.092 $m^2\ V^{-1}\ s^{-1}$. The λ_L value of PbTe is 2.0 W $m^{-1}\ K^{-1}$ at room temperature.

The z value of PbTe is only just over 1 x $10^{-3}\ K^{-1}$ at room temperature [5.79], but at higher temperatures, it eventually becomes superior to that of Bi_2Te_3. It is preferable to use solid solutions of PbTe rather than the simple compound. Alloys can be formed by substituting tin for lead or by substituting sulfur or selenium for tellurium.

Airapetyants et al. [5.36] carried out experiments on both the $PbTe_xSe_{1-x}$ system and the $Pb_{1-y}Sn_yTe$ system. They suggested that the former was more useful for p-type thermoelements because one effect of alloying was to reduce μ_n, whereas the latter would be the better n-type material because it displayed a reduced μ_p. Later work indicates that $Pb_{1-y}Sn_yTe$ is suitable for both n-type and p-type thermoelements and that the reduction in λ_L more than offsets any fall in μ [5.80]. There is, however, a problem with $Pb_{1-y}Sn_yTe$; the energy gap becomes zero when y is equal to about 0.4. Thus y should not be allowed to become too large; Rosi et al. [5.81] recommended the composition $Pb_{0.75}Sn_{0.25}Te$.

It is possible to obtain a value of z_nT near unity for materials based on lead telluride at a temperature of about 700 K, but the value of z_pT is only about 0.7 However, a value of z_pT in excess of unity can be obtained by using a so-called TAGS formulation.

The acronym TAGS stands for alloys containing elements Te, Ag, Ge, and Sb. They are essentially alloys between the compounds $AgSbTe_2$ and GeTe and are closely related to lead telluride because part of the solid solution range has the same sodium chloride structure. However, there is a transformation to a rhombohedral structure when the concentration of GeTe becomes less than about 80% and, in fact, the highest figure of merit is displayed close to the phase transformation. It is believed that lattice strains associated with this transformation may be effective in reducing the lattice thermal conductivity and, thus, in increasing the figure of merit, as shown by Skrabek and Trimmer [5.82].

Skrabek and Trimmer examined the variation of thermal conductivity with composition for concentrations of GeTe between 50% and 90%. There was a smooth variation between 50% and about 75% GeTe which gave values that lined up with the thermal conductivity of the alloys containing more than 90% GeTe. However, between 75% and 90% GeTe, there were distinctly smaller thermal conductivity values. Moreover, the Seebeck coefficient and the electrical conductivity were both higher than expected from the data outside this range. Obviously, then, the figure of merit is the highest within the range of compositions that corresponds to the changeover from a rhombohedral to a rock salt structure.

The highest figure of merit was observed for so-called TAGS-80, the composition with 80% GeTe. However, the mechanical strength of this material is inferior to that of TAGS-85, 85% GeTe. This may be associated with the fact that there is a minimum in the coefficient of thermal expansion at the 85% composition, which also corresponds to a minimum lattice parameter. The lattice conductivity of TAGS-85 is compared with that of lead telluride in Fig. 5.27, and this in itself explains the greatly improved figure of merit.

It has been noted that some of the nonstoichiometric alloys, at any given GeTe content, display increased values for the power factor but, because of corresponding increases in the thermal conductivity, there is no improvement in the figure of merit. TAGS-85 actually gives a value for zT of about 1.4 at a temperature of 750 K. As the temperature is lowered, zT becomes smaller but remains above 0.8 for temperatures above 500 K. In this region, the Seebeck coefficient lies between 150 and 200 $\mu V K^{-1}$.

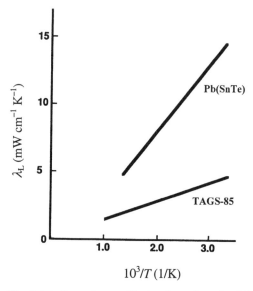

Fig. 5.27. Dependence of lattice thermal conductivity on temperature for TAGS-85 and lead telluride, from Skrabek and Trimmer [5.82]

Even higher values for z have been reported for the AgSbTe$_2$ alloy that contains 80% GeTe [5.83]. This material is claimed to be superior to TAGS-85 over the complete temperature range, and zT is equal to no less than 1.7 at a temperature of 700 K. Whether this alloy is as good as TAGS-85, when stability and strength are taken into account, remains to be seen, but there is no doubt that, within their particular temperature regime, the alloys of AgSbTe$_2$ with GeTe outperform any other thermoelectric materials.

5.3.2 Silicon–Germanium Alloys

Neither silicon nor germanium has a particularly large z at any temperature because, even though both materials have high carrier mobilities, they possess very large values of λ_L. However, λ_L is considerably reduced when solid solutions between the two elements are formed, and these silicon–germanium alloys are of particular interest for thermoelectric applications.

Silicon and germanium form a complete range of solid solutions. For any of the compositions, the solidus temperature lies between the melting points of the elements, 1693 K and 1231 K, respectively. A typical thermoelectric alloy has the composition Si$_{0.7}$Ge$_{0.3}$ and a solidus temperature of about 1500 K. This alloy shows negligible degeneration when operated at a source temperature of 1300 K [5.84]. The energy gaps for silicon and germanium are 1.15 eV and 0.65 eV, respectively, so the alloys at the silicon-rich end have only a small concentration of minority carriers when they are optimally doped for thermoelectric generation, even at high temperatures.

The lattice thermal conductivities of silicon and germanium are approximately 113 and 63 W m^{-1} K^{-1}, respectively, at 300 K, but for Si$_{0.7}$Ge$_{0.3}$, the value is about 10 W m^{-1} K^{-1} [5.85]. It is remarkable that this reduction of λ_L is accompanied by very little, if any, reduction in μ_n and μ_p below those of elemental silicon. Thus, zT for either p-type or n-type material is of the order of 0.5 or greater at temperatures above 600 K [5.79].

The considerable reduction of λ_L of silicon–germanium alloys has an interesting consequence. The scattering of high-frequency phonons in the alloy system is strong enough to make λ_L particularly sensitive to grain boundary scattering. Thus, in spite of the relatively large mean free path lengths for the charge carriers, silicon–germanium alloys are the only materials in which it definitely appears that a reduction in grain size for polycrystalline materials is beneficial [5.86]. Note that the crystal structure in the silicon–germanium system is that of diamond, which is cubic, and there is, in principle, no disadvantage in using polycrystalline samples. However, Slack and Hussein [5.87] suggest that the improvement in the figure of merit for fine-grained silicon–germanium will be insignificant.

Slack and Hussein [5.87] carried out a comprehensive review of silicon–germanium and attempted to determine an upper bound for z that might be achieved for these alloys. These authors concentrated their attention on Si$_{0.7}$Ge$_{0.3}$, which is the preferred composition. They noted that practical generators based on silicon–germanium alloys display an efficiency of 6% when they operate between a source and sink at 1000°C and 300°C, respectively, although optimized silicon–germanium couples might be twice as efficient. Their object was to determine the effect of decreasing the lattice conductivity, if this could be accomplished without changing the electronic parameters.

The electronic properties of silicon and germanium have been widely studied in view of the importance of these elements in the development of the microelectronics industry. Even though these studies have been directed mainly toward materials of low carrier concentration at modest temperatures, it is still possible to predict what will happen at the high doping levels and high temperatures that are characteristic of thermoelectric generation. The solid solutions have not been as thoroughly studied as the elements, but, nevertheless, there is also a wealth of experimental data on these materials.

The basis of the calculations is the established band structure. At the doping levels in which we are interested, the complexities of the hole band near its edge are unimportant. However, when we consider the electrons, we must take account of two sets of valleys with their extrema at different levels. These sets are known as the X and L bands, respectively. Slack and Hussein expressed the Seebeck coefficient and electrical conductivity in terms of the band parameters and mobilities for different carrier concentrations at each temperature. Consequently, the dependence of the power factor on the doping level could be found. Because of complexities in the band structure, the power factor in the model calculation for n-type material exhibits two maxima. One is associated with the X band and lies close to a carrier concentration of 2×10^{26} m^{-3}: the other is related to the L band and lies near a concentration of 2×10^{27} m^{-3}. At high temperatures, there is not much difference between the two maxima so far as power factor is concerned, but one must remember that the electronic thermal conductivity also has an effect on the figure of merit. Thus, it might seem that one should choose n-type material with a carrier concentration close to 2×10^{26} m^{-3}. This is borne out when the figure of merit is calculated below 1100 K, but at higher temperatures, the maximum in the power factor at the higher carrier concentration is sufficiently great for this to yield the better figure of merit. Similar complications do not occur for the p-type Si–Ge alloys.

Slack and Hussein [5.84] found that their predicted values for the power factor were in good agreement with the experimental data. Thus, they were encouraged to proceed with their calculations using both the known values of λ for large pure crystals and their predicted values when additional forms of phonon scattering are introduced. They introduced a parameter f which takes the value zero when phonon–phonon and alloy scattering are present and the value unity when the so-called minimum lattice conductivity is reached in amorphous material.

When $f = 0$, the figure of merit has the value that is typical of optimized large crystals. For n-type Si$_{0.7}$Ge$_{0.3}$, zT varies from 0.3 at 500 K to 1.1 at 1200 K. As f becomes larger, the figure of merit rises, and when the material is fully amorphous, the value of zT is expected to range from about 1.0 at 500 K to about 1.8 at 1200 K. Of course, the magnitude of the optimum Seebeck coefficient becomes greater as the lattice conductivity is reduced, because then it is necessary to lower the electronic thermal conductivity. It goes without saying that this calculation is based on the assumption that f can be made larger than zero without affecting the electronic properties and, so far, this has not been achieved. It has already been mentioned that Slack and Hussein [5.87] do not think that the boundary scattering of the phonons will improve the figure of merit. Their argument is based on a belief that boundary resistance will affect the electrical conductivity more than was assumed by Rowe and Bhandari [5.82]. However, they suggest that introducing of a finely dispersed second phase may be more helpful.

5.4 Production of Materials

We have seen that both bismuth telluride alloys and bismuth–antimony alloys have preferred crystallographic directions for thermoelectric and thermomagnetic applications, so that single crystals or, perhaps, aligned polycrystalline materials give the best performance. There is also usually the need to add the appropriate amounts of donor or acceptor impurities to yield the optimum Seebeck coefficient or electrical conductivity. Furthermore, the concentrations of these impurities and the proportions of the main constituents of any solid solutions must be controlled within appropriate limits. Thus, in this section, we discuss the techniques for producing good thermoelectric materials. Although we shall dwell mainly on the preparation of bismuth telluride alloys, the principles are applicable to the other materials that have been mentioned.

Note that Ioffe [5.75] has considered the effect on the figure of merit of a statistical departure of the electrical conductivity from its optimum value. Perhaps of greater importance is the difficulty that is experienced by module manufacturers if the resistance of thermoelements is different from the designed value because this means that the current for a given cooling power is altered. It has also been claimed that internal circulating currents, which arise from nonuniformity of thermoelectric coefficients, increase heat conduction losses [5.88]. There is no doubt that the best results have always been achieved with properly homogenized material.

We shall discuss the various techniques that have been used in producing thermoelectric materials and shall illustrate them with special reference to bismuth-telluride and its alloys and bismuth–antimony alloys.

5.4.1 Segregation and Constitutional Supercooling

The best thermoelectric and thermomagnetic materials are usually grown from a melt. Therefore, we discuss the problems that arise from the segregation of constituents with the advance of the solid–liquid interface. Unless this process is understood, it is unlikely that the solid will have the required composition, and it may well be inhomogeneous on both macroscopic and microscopic scales.

Let us consider the equilibrium phase diagram for either an element or a compound with a very narrow range of composition, to which a soluble impurity is added. This impurity may either raise or lower the melting temperature, as shown in Fig. 5.28 (a) or (b). The liquid and solid regions in the diagrams are separated by an intermediate region in which the liquid and solid phases are both present. Suppose that we allow the temperature of the liquid, at a certain composition, to fall slowly until the liquidus line is reached at point A. If it is assumed that the system remains in equilibrium, further removal of heat will result in solidification of some of the charge, and the composition of the solid will correspond to point B in the phase diagram. This, of course, means that the composition of the remaining liquid will change from that at point A, in a direction opposite to that of point B. In particular, when the segregation coefficient k is greater than unity, the impurity concentration in the liquid will gradually become smaller, and, when $k < 1$, the impurity content in the liquid will gradually increase. The segregation coefficient is defined as the ratio between the impurity concentrations at points B and A.

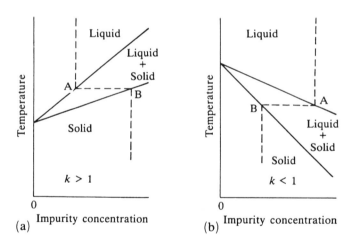

Fig. 5.28. Equilibrium phase diagrams showing schematically the liquidus and solidus lines as a function of impurity concentration for the segregation coefficient (**a**) greater than and (**b**) less than unity

The process of segregation is used in purifying materials. Thus, in directional-freezing, one end of the melt is allowed to solidify first, whereupon the liquid–solid interface moves continuously to the other end. When $k < 1$, the first end to freeze becomes relatively pure and the impurities segregated into the melt distribute themselves preferentially near the other end.

The effectiveness of segregation for purification was greatly improved by Pfann [5.89] when he developed his zone-refining technique. In this method, a molten zone is made to traverse a solid ingot. When $k < 1$, the impurities tend to be carried along with the liquid. The material becomes even purer when more than one zone is passed; it is typical for a number of zones to be passed. Zone-melting techniques can be used to produce uniform materials of specified impurity content.

So far, it has been assumed that the liquid–solid interface moves slowly enough to maintain equilibrium. Now, we consider what happens when growth proceeds much more rapidly. In this case, the composition of the liquid is no longer uniform and, because of the segregation effect, there will be a transient region adjacent to the solid that has a reduced melting temperature. Reference to Fig. 5.28 shows that the effect of segregation is to reduce the liquidus temperature in the transient region, irrespective of whether k is greater than or less than unity. The way in which this temperature varies with distance from the interface is shown schematically in Fig. 5.29, where the width of the transient region is expected to depend on the rate at which the interface is moving and on the diffusion coefficient of the impurities in the liquid.

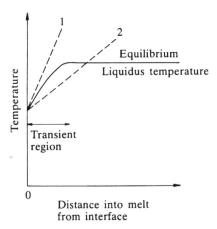

Fig. 5.29. Temperature distribution in the melt near an advancing solid–liquid interface. The solid lines show the equilibrium liquidus temperature, and the broken lines are examples of the actual temperature in the melt, as discussed in the text

The broken lines in Fig. 5.29 represent the actual temperature distribution in two typical cases. Case 1 does not lead to any special problem, but in case 2 the temperature gradient is less than the gradient of the liquidus temperature, where so-called constitutional supercooling occurs. It will be seen that there is a tendency for freezing to occur in advance of the interface. What happens in practice is that solidified cells form in advance of the main front, and liquid of lower melting temperature is trapped between the cells. When constitutional supercooling has occurred, the resultant ingot is inhomogeneous on a microscopic scale and has inferior thermoelectric properties.

The conditions for avoiding constitutional supercooling were first specified by Tiller et al. [5.90], who assumed that the melt was unstirred. It might be thought that stirring would reduce the width of the transient region, and indeed it does, but a boundary layer with a compositional gradient always remains in the liquid. As Hurle showed [5.91], the condition for constitutional supercooling remains independent of the stirring rate for a specified composition of the solid. According to Hurle, constitutional supercooling is avoided when the temperature gradient in the melt satisfies the inequality, where v is the rate of advancement of the interface, L is the latent heat, ρ_d is the density:

$$\frac{dT}{dx} > v\left(\frac{L\rho_d}{\lambda_S} - \frac{mC_S \lambda_L (1-k)}{D\lambda_S k}\right), \tag{5.9}$$

m is the slope of the liquidus line, C_S is the concentration of solute in the solid, and D is its diffusion coefficient in the liquid. λ_L and λ_S are the thermal conductivities of the liquid and solid, respectively. If we assume that the first term on the right-hand side of the inequality (5.9) is negligible compared with the second and if we suppose that λ_L and λ_S are approximately equal near the melting temperature, we find the much simpler expression,

$$\frac{dT}{dx} > \frac{v\Delta T}{D}, \tag{5.10}$$

where ΔT is the temperature difference between the liquidus and solidus lines at the composition in which we are interested. The inequality illustrated in (5.10) makes it clear that constitutional supercooling can be prevented only by employing a sufficiently large temperature gradient or a sufficiently small growth rate because ΔT and D cannot be altered.

5.4.2 Growth From a Melt

The alignment that is preferred for such materials as Bi_2Te_3 and bismuth–antimony is best achieved by growing from a melt. In Bi_2Te_3, for example, there is a strong tendency for the cleavage planes to lie perpendicular to the liquid–solid interface [5.28]. The growth direction is the preferred direction for current flow in the thermoelements.

Both aligned polycrystalline ingots and single crystals of bismuth telluride alloys have been produced by vertical and horizontal directional freezing and zone melting. Whether or not a single crystal results depends on the shape of the growth front, as well as on the steps that have been taken to provide a dominant crystal at the first end to freeze. A near planar interface, or one that is convex toward the liquid, is desirable if crystals are not to grow from the containing walls. Of course, aligned polycrystals may be preferred to single crystals because the latter are more fragile.

The pseudobinary phase diagram for alloys between Bi_2Te_3 and Sb_2Te_3 is shown in Fig. 5.30 (a) [5.92]. It appears that the liquidus and solidus curves come together at concentrations of about 33% and 67% Sb_2Te_3. Even at other compositions, the two curves are not far apart. Thus, segregation of the major components is unlikely to be a serious problem in bismuth–antimony telluride.

Figure 5.30 (b) shows the pseudobinary phase diagram for the bismuth seleno-telluride system [5.93]. Here the liquidus and solidus curves are more widely separated, but segregation of the major components should not worry us too much because the preferred alloys contain no more than about 10% Bi_2Se_3. Under this

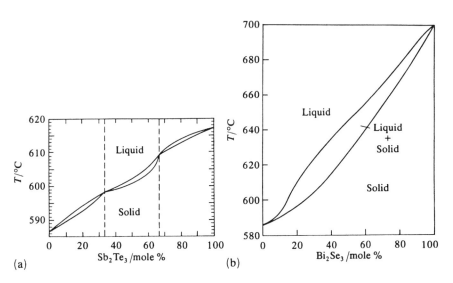

Fig. 5.30. Pseudobinary phase diagrams for **(a)** the bismuth–antimony telluride system and **(b)** bismuth selenotelluride alloys

condition, the segregation coefficient is not too far from unity. Segregation of the doping agents is a much more serious matter, and for this reason, care must be taken to avoid the effects of constitutional supercooling. The influence of the growth rate and the temperature gradient in the Bridgman technique was first studied by Cosgrove et al. [5.88], and the results have since been confirmed by other workers [5.94].

Cosgrove and his colleagues arranged for furnace gradients of 2.5, 5.0, or 25 K mm^{-1}, and growth rates between 2.2×10^{-4} and 4.2×10^{-2} mm s^{-1} were employed. Particular attention was paid to studying the increase of λ due to circulating thermoelectric currents in material that displayed microscopic inhomogeneity. The calculated electronic component was subtracted from the total measured thermal conductivity to yield a quantity that must be the sum of λ_L and a contribution, λ_{circ}, from the circulating currents. The variation of this quantity with the ratio of the rate of growth v to the furnace gradient dT/dx is shown in Fig. 5.31. Although the data are somewhat scattered, it is quite clear that there is an undesirable increase in the apparent λ at high growth rates and low temperature gradients. The results suggest that the ratio $v(dT/dx)^{-1}$ should be somewhat less than 5×10^{-4} mm^2 K^{-1} s^{-1} to produce a good thermoelectric material. The required condition is likely to depend on the alloy composition and the doping agent that is employed, as shown in Vol'pyan et al. [5.95]. Note that a more recent study of circulating thermoelectric currents suggests that they are not the reason for the behavior that was observed by Cosgrove and his co-workers [5.91]. However, this does not alter the fact that the growth conditions should be such as to avoid constitutional supercooling.

The procedure described by Horst and Williams [5.94] is typical of that which yields high-quality bismuth–telluride alloys. They used starting materials of 99.9999%

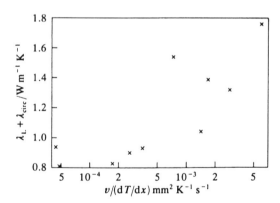

Fig. 5.31. The sum of the lattice conductivity λ_L and the thermal conductivity λ_{circ} due to circulating currents in BiSbTe$_3$, plotted against the ratio of the growth rate v to the furnace gradient dT/dx

purity. Less pure materials are satisfactory if one is prepared to compensate for any excess donors or acceptors that might be present. Their elements were first melted in a quartz capsule under hydrogen. After cooling, the capsule was sealed with a residual hydrogen pressure of about 10 Pa. The charge was held in a vertical furnace above the melting temperature for 24 hours and then lowered through a temperature gradient of 4 K mm^{-1} to solidify at the rate of 5 mm hr^{-1}.

Successful crystal pulling of Bi_2Te_3 has been described by Yang and Shepherd [5.96] and Strieter [5.97]. Pulled crystals are suitable for basic measurements of the transport properties but are less desirable when controlled concentrations of dopants are needed, as in optimized thermoelements.

Now, we turn to the production of bismuth–antimony. Here, one generally requires single crystals, although, for thermoelectric applications in the absence of a magnetic field, it may be satisfactory to use polycrystals in which the c axes are closely parallel. Crystals of bismuth can be pulled from the melt [5.98], but the alloys present special problems, as indicated by the phase diagram in Fig. 5.32. The liquidus and solidus curves are widely separated, even for small concentrations of antimony in bismuth, so that the ratio of the speed of growth to the temperature gradient must be very small if constitutional supercooling is not to occur. Furthermore, because the melting temperature is relatively low at the bismuth-rich end of the system, it is difficult in practice to achieve very steep gradients during melt growth. Quite apart from its contribution to difficulties of microsegregation, the large value of the segregation coefficient must be taken into account when the question of macroscopic uniformity is considered. The concentration of antimony in the liquid is only about one-tenth of that in the crystallizing solid.

It may be best to use a horizontal zone-melting technique to produce bismuth–antimony alloys. Uher [5.99] prepared crystals containing up to 10 at.% of antimony by such a method. His technique typifies what is needed for successful growth. A charge of the appropriate composition was first melted under hydrogen in a sealed

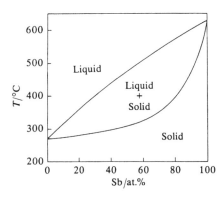

Fig. 5.32. Phase diagram for the Bi–Sb system, from Hansen [5.120]

tube and then quenched. It was then placed in a horizontal quartz boat contained in a quartz tube under hydrogen, as shown in Fig. 5.33. A feature of Uher's apparatus was the use of tubular quartz heaters with water-cooled metal reflectors to concentrate the radiation over a narrow region. Although a water-cooled jacket was provided near the melting interface to limit the zone width, it was found that too steep a gradient led to a lineage structure in the crystal. In fact, it was beneficial to use afterheaters to limit the gradient to about 3 K mm^{-1}. The inside of the boat was coated with graphite to provide a soft seating for the charge. A zone-leveling process, consisting of several high-speed forward and reverse passes, was followed by a final pass at a very slow speed of the order of 1 mm hr^{-1}.

A slow growth rate requires strict control of the ambient temperature and of the heater power. It is also necessary to maintain the mechanical drive at a constant speed. Uher's equipment was mounted on a lathe bed resting in sandboxes in a temperature-controlled room, and the lathe drive was employed to move the entire heater assembly, through reduction gears, at the required rate. It was found that uniform crystals were produced when the ratio of the temperature gradient to the growth rate lay above the curve shown in Fig. 5.34, which is based on the inequality in (5.10). We use the estimate by Brown and Heumann [5.100] for the liquid diffusion coefficient D of bismuth–antimony, that lies between 0.2 and 0.3 mm^2 s^{-1} at 300°C, and rises to about 0.9 mm^2 s^{-1} at 700°C.

The validity of the condition specified by the curve in Fig. 5.34 was first established by Brown and Heumann. The observations by Yim and Amith [5.45] on $Bi_{0.95}Sb_{0.5}$ were consistent with the same criterion. Short and Schott [5.101] demonstrated the strong dependence of the so-called resistivity ratio (ρ at 4.2 K divided by ρ at 300 K) on the rate of growth, and it may be inferred that the thermogalvanomagnetic properties in general are markedly affected by microsegregation.

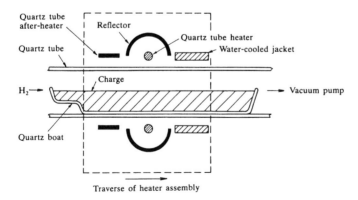

Fig. 5.33. Horizontal zone-melting apparatus used by Uher [5.99] for bismuth–antimony

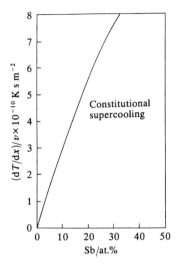

Fig. 5.34. The boundary condition for constitutional supercooling in the Bi–Sb system. Micro-segregation is avoided if $(dT/dx)v^{-1}$ lies above the curve [5.99]

The results obtained from the growth of both bismuth telluride alloys and bismuth–antimony alloys confirm Hurle's theory of constitutional supercooling. The same principles can, of course, be applied to other thermoelectric materials.

5.4.3 Powder Metallurgy

Although there is no doubt that the highest figures of merit are usually achieved with materials that are grown from a melt, there are certain advantages in using a powder metallurgy technique. Sintered materials are invariably more robust than those that are melt-grown, and they do not require the same precautions to establish homogeneity. They also offer the possibility of improving z by reducing λ_L through scattering phonons on the grain boundaries. However, sintered thermoelectric materials are satisfactory only if the density is reasonably high and if there is negligible grain boundary electronic resistance. In addition, some degree of alignment of the grains is desirable for materials in which the thermoelectric properties are anisotropic.

The principles of powder metallurgy may be illustrated with reference to Bi_2Te_3 and its alloys. A typical procedure is shown in Fig. 5.35. The major elemental components and any required impurities are first melted together in a sealed quartz ampoule and then quenched. The casting is then powdered by rod or ball milling or some other comminution process. It is usual to select a specific range of powder sizes. Therefore, the powder must be screened using calibrated sieves. It is then

160   5. Review of Established Materials and Devices

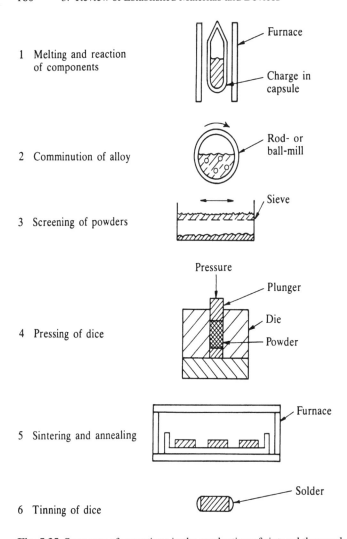

1 Melting and reaction of components

2 Comminution of alloy

3 Screening of powders

4 Pressing of dice

5 Sintering and annealing

6 Tinning of dice

Fig. 5.35 Sequence of operations in the production of sintered thermoelements

compressed in a die to form a pellet of more or less the required dimensions. The green-state pellet is heat treated to sinter the particles together and to anneal the material. The changes in dimensions during heat treatment should be effective in producing a thermoelement of precisely the required size. Finally, the ends of the sample have to be tinned with an appropriate solder, probably after plating.

Successful production of thermoelements has been carried out using a wide range of conditions. For example, Cope and Penn [5.102] used particle sizes from 150 μm up to 1mm and concluded that the main effect of reducing the particle size lay in the increased risk of atmospheric contamination. Oxygen degrades the thermoelectric properties [5.103]. Therefore, it is important to reject the finest particles and to prevent the remaining powder from contact with the atmosphere until it has been pressed. Durst et al. [5.104] showed that particles in the range of 100 to 250 μm should not be too susceptible to atmospheric contamination, and at the same time are small compared with the smallest dimension of a thermoelement. This is significant because large grains in a sintered sample can impair its mechanical properties.

In a cold-pressing process, Cope and Penn [5.102] found that the density of sintered samples was highest when the compacting pressure was 5×10^5 kPa, although the lower density specimens made by Durst et al. [5.104] had an equally good z. The σ of material, that has a porosity δ is reduced to the value $\sigma(2 + \delta)/2(1-\delta)$, but, because λ is reduced by the same factor, z is not affected. Likewise, the choice of sintering temperature is not too critical, provided that an appropriate doping level is selected. The sintering temperature should not lie below 300°C, or the sintering process is too slow; temperatures above 450°C lead to distortion and swelling. This still leaves a fairly wide range. Typically, samples may be sintered in an inert or reducing atmosphere for a period of an hour or more at a temperature of 400°C.

Cold-pressed Bi_2Te_3 alloys do not usually exhibit any substantial alignment of grains. However, some measure of alignment, where the basal planes are normal to the direction of pressing, generally results from a hot-pressing process, such as that described by Airapetyants and Efimova [5.105]. These workers used a pressure of 7×10^5 kPa at a temperature of 400°C and a subsequent heat treatment at 350°C.

Note that even cold-pressed sintered thermoelements have a higher ρ in the direction of pressing, compared with that in a perpendicular direction. This is partly due to so-called pressing faults, that is, cracks of microscopic proportions. These thermoelements should be used with current flow perpendicular to the pressing direction. Whether true preferential alignment exists is best determined by a technique that does not depend on the current flow. Goldsmid and Underwood [5.106] used an X-ray method and also a technique that was based on a determination of the anisotropy of the Seebeck coefficient at temperatures that were high enough for mixed or intrinsic conduction to be displayed. If there is no preferential alignment, z of the p-type alloys of Bi_2Te_3 is reduced by 6% and that of the n-type alloys by 20%, compared with values for properly oriented single crystals.

Sintered bismuth–antimony alloys are much preferable to single crystals from the viewpoint of mechanical strength but have very much reduced values for both z and z_E. An attempt to improve the figures of merit by using a small grain size failed because the reduction in λ_L by boundary scattering was more than offset by a reduction in μ [5.107]. Nevertheless, it is most desirable to find some substitute for single crystals of bismuth–antimony because the requirement that current flows in the c direction makes the thermoelements particularly susceptible to mechanical damage.

Now, it seems that a reasonable compromise can be achieved by using bismuth–antimony that is extruded and annealed [5.108]. Likewise, extrusion can be used to produce Bi_2Te_3 alloys with good thermoelectric properties [5.109]. There is, however, still some degradation of performance compared with the best that is observed for melt-grown material.

Thermoelectric generator materials can also be prepared by using powder metallurgy techniques. Most of them are made from crystals that have cubic symmetry and there is, therefore, no advantage to be gained from aligning the grains. Moreover, in the case of silicon–germanium alloys, the reduction of grain size may lead to an improvement in z through boundary scattering of the phonons [5.86]. It is clear that sintering and allied techniques will continue to be important for producing thermoelectric materials.

5.4.4. Thin and Thick Films

At present, thermoelements in the form of films are used only in rather specialized applications. This is not necessarily because Z is lower for laminar thermocouples but because of the severe heat transfer problems associated with such devices. If heat flow is perpendicular to the surface, then a given temperature difference implies a very large heat flux. On the other hand, heat flow parallel to the surface is accompanied by heat losses through the substrate. Nevertheless, economy in the use of materials could result from using very small thermoelements. There is also ongoing research to enhance z from quantum-well effects that may require producing the materials as films. Thus, a brief mention of thin and thick film technology, applied to thermoelectric materials, is in order. In Part II of the text, we will discuss the research into quantum confinement effects, as well as methods for synthesizing thin films.

Thermoelements are invariably made from alloys or compounds, and a major problem is controlling the relative concentrations of the components. When thin films are being made, this is accomplished more readily using a flash-evaporation technique rather than coevaporation of the components from different sources. Slow evaporation from a single source is rarely successful because of the different vapor pressures of the constituents. Flash evaporation of n-type Bi_2Te_3 alloys has been carried out successfully by Nakagira et al. [5.110], as evident from the high value for the power factor $\alpha^2\sigma$ that they reported. The p-type material produced by the same method had very poor properties, but Nakagira and his colleagues found that they could make good p-type thick films by using a casting technique. Therefore, conventional refrigerator alloys can be produced, in one laminar form or another, with properties that are comparable with those of bulk materials. In general, it is necessary to lay down the films and also to anneal them until μ reaches its crystalline value.

5.5 Design of Modules

5.5.1 Single-Stage Refrigerators and Generators

Earlier in the text, we considered the factors that control the coefficient of performance ϕ_{max} of a thermoelectric refrigerator. Now, we shall discuss the problem of obtaining a specific cooling power. It is rare that a single thermocouple can be used for our purposes; instead one must use what is commonly called a thermoelectric module, consisting of a multiplicity of thermoelements. Similar problems are encountered in designing thermoelectric generators.

If Z is independent of temperature, one may use one of the curves in Fig. 1.7 to determine ϕ_{max}. More generally, empirical curves similar to those in Fig. 1.7 can be drawn for a specific pair of thermoelectric materials after their properties have been measured as a function of temperature. Knowing ϕ_{max} and the required cooling power NQ_c, where N is the number of thermocouples in the module (as yet unspecified) and Q_c is the cooling power of each couple, one may determine the required electrical power NW which is equal to NQ_c/ϕ_{max}.

In principle, one may select N quite arbitrarily, but practical considerations must be taken into account. For $N=1$, it is usual that an extremely large electric current would be required. There are also problems associated with thermal expansion when one attempts to use thermoelements of very large cross-sectional area. It is usual to set a limit on the current, and N is made large enough to ensure that this current is not exceeded. The required resistance R of each couple can be determined from empirical data. The power input W per couple can be obtained from (1.19). Because NW is already known, the lower limit for N can be determined. In practice, one would probably select a larger value of N, so that a regular array of thermoelements could be formed.

We may illustrate these principles by considering a specific example. Suppose that the thermocouple materials have z equal to 2.5×10^{-3} K^{-1} and that the hot and cold junctions are to be at 300 K and 260 K, respectively. Suppose, also, that cooling power of 10 W is needed for our particular application. Then, ϕ_{max} is equal to 0.423, and NW must be 23.63 W. Now, let us assume that the current must not exceed 5A. The resistance of each couple should be at least 10.5 mΩ, assuming that $(\alpha_p - \alpha_n)$ for our pair of materials is equal to 400 μV K^{-1}. From (1.19), we find that the electric power per couple is 0.3425 W. The minimum number of couples turns out to be seventy. By chance, it happens that the module could be formed from a regular 14 x 10 matrix of thermoelements.

From the users viewpoint the problem is somewhat different. One has to decide which of a manufacturer's modules will be most suitable for a given task. The procedure has been outlined by Buist [5.111] and will not be recapitulated here. It should be remembered that a cooling unit may incorporate more than one module and that a combination of two or more in parallel can sometimes be more suitable than a single large module.

Nowadays, most thermoelectric modules incorporate the thermoelements and the metal connectors and also end plates that allow the transfer of heat but not electricity. A typical form of construction is shown in Fig. 5.36. The thermoelements are connected by copper links that have been screen printed on sheets of either alumina or beryllia. Alumina is a reasonably good conductor of heat, and beryllia is even better. The screen-printed copper on the exterior faces of the ceramic sheets is used to attach the module to the source or sink. The printed copper that forms the links must be thick enough to avoid adding any appreciable electrical resistance.

5.5.2 Heat Transfer Problems

The elementary theory of thermoelectric refrigeration ignores two heat transfer problems that are invariably encountered in practice. First, there is always some transfer of heat between the hot and cold junctions through the space between and around the thermoelements. Second, there is thermal resistance between the hot and cold junctions of the module and the sink and source, respectively. The two factors are related to one another because a reduction of the space between the thermoelements can lead to an increase in the thermal resistance at the interfaces, as the total cross-section becomes smaller.

We examine the problem with reference to the model shown in Fig. 5.37. It is supposed that the space between the thermoelements has a cross-sectional area that is a fraction g of that of the elements themselves and that this space is filled with a material whose thermal conductivity is λ_1. The value of λ_1 should, of course, be as small as possible. Although we discuss a single couple, our considerations apply equally well to a multijunction module.

The effect of heat transfer through the thermal insulation is to decrease the thermal resistance by a factor of $(1 + \lambda_1 g/\lambda)$ where λ is assumed to be the same for both branches. The z value is reduced by this factor.

Now, let us suppose that the electrical insulation between the thermoelectric device and the source and sink has a thermal conductance k_c per unit area. Thus, if the couple itself has a cross-sectional area A, the total cross-section is $A(1 + g)$, and the

Fig. 5.36. Cross-section through a module that has screen-printed copper links on insulating ceramic plates

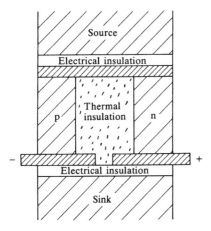

Fig. 5.37. Theoretical model of a refrigerating thermocouple incorporating heat transfer around the thermoelements and thermal resistance beyond the junctions

thermal conductance at each interface is $k_c A(1 + g)$. When the cooling power is Q_c and the coefficient of performance is ϕ, the rate at which heat is delivered to the sink is $Q_c(1 + 1/\phi)$. At the interfaces with the source and sink, there will be unwanted temperature differences equal to $Q_c / k_c A(1+g)$ and $Q_c(1+1/\phi) / k_c A(1+g)$, respectively. Then, the temperature difference ΔT across the thermocouple is related to the difference ΔT^* between the source and sink by

$$\Delta T = \Delta T^* + \frac{Q_c(2+1/\phi)}{k_c A(1+g)}. \tag{5.11}$$

Because $\Delta T > \Delta T^*$, the coefficient of performance of the device is reduced.

If the length L of the thermoelements could be made large enough, g could be reduced to zero and, at the same time, there would be very little difference between ΔT and ΔT^* because A would also be very large. Practical considerations restrict L to no more than a few millimeters. We discuss, then, selecting g when L is specified.

To some extent, the solution of the problem depends on the condition under which the device is to operate. One may, for example, be interested in obtaining the maximum coefficient of performance or perhaps the maximum cooling power. Here we consider selecting g to make the maximum temperature difference at zero load as large as possible.

As the temperature difference approaches its maximum value, $Q_c(2+1/\phi)$ approaches the electrical power W that is supplied to the device. If there were no heat transfer losses or thermal resistances at the interfaces, the maximum temperature difference would be equal to $ZT_C^2/2$, as indicated in Table 1.2. The input power would be equal to $(\alpha_p-\alpha_n)^2 T_C T_H/R$, according to Table 1.2. For convenience, we may write W as $\lambda A Z T_C T_H / L$, and then we find that

$$\Delta T_{max} = \Delta T^*_{max} + \frac{\lambda Z T_C T_H}{k_c L(1+g)}. \tag{5.12}$$

If we take into account the reduction of the effective figure of merit due to heat transfer through the insulation,

$$\Delta T_{max} = \frac{ZT_C^2}{2(1+\lambda_1 g/\lambda)}. \tag{5.13}$$

Thus, combining (5.12) and (5.13), we obtain

$$\Delta T^*_{max} = \frac{1}{2} Z T_C^2 \left\{ \frac{1}{1+\lambda_1 g/\lambda} - \frac{2\lambda T_H}{k_c L(1+g) T_C} \right\}. \tag{5.14}$$

It is our objective to make the expression inside the brackets as close to unity as possible by an appropriate choice of g. The optimum value of g will depend on the quality of the insulation between the thermoelements and the effectiveness, as a thermal conductor, of the electrical insulation at the source and sink. The use of (5.14) is given for a typical case. We suppose that ZT of the thermocouple is close to unity so that T_H/T_C is about 1.4. The δ value of the thermoelectric materials will be about 1.5 W m^{-1} K^{-1}. It will also be supposed that the space between the elements is filled with expanded polyurethane and λ is equal to 0.02 W m^{-1} K^{-1}. The electrical insulation at the interfaces with the source and sink it will be assumed, consists of greased polytetrafluoroethylene (PTFE) film about 1.3 μm thick, which has a thermal conductance per unit area of 400 W m^{-2} K^{-1}.

In Fig. 5.38, the ratio $\Delta T^*_{max} / (ZT_C^2/2)$ is plotted against the length of the thermoelements for various values of g. Note that if $L < 10$ mm, ΔT^*_{max} never rises above 94% of the value that it would have for an ideal couple. Commercial modules commonly employ thermoelements only 2 or 3 mm long, and, with the chosen parameters, λ_1 and k_c, the maximum temperature difference would be less than 90% of the ideal value. In practice, present-day modules use thin sheets of beryllia or alumina ceramic, permanently bonded to the thermocouple junctions, to provide the necessary electrical insulation. Such sheets have improved values for k_c allowing the optimum value of g to be reduced and the maximum temperature difference to approach the ideal value more closely.

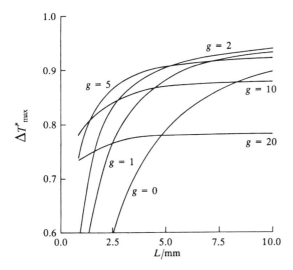

Fig. 5.38. Plots of the ratio of the maximum temperature difference, $\Delta T^*_{\max}$, to $ZT_c^2/2$ against the length L of thermoelements for various values of g, where $(1+g)^{-1}$ is the proportion of the total cross-sectional area that is occupied by the thermoelements. The design parameters are given in the text

5.5.3 Contact Resistance

Another problem that can become important when the length of the thermoelements is small is the electrical resistance of the contacts with the metallic connectors or, indeed, of the connectors themselves. Resistive contacts increase the electrical power consumption and, add to the thermal load at the cold junction. Here, we assume that the contact resistance is equally distributed between the hot and cold junctions, and we follow the treatment of Parrott and Penn [5.112].

Contact resistance is usually inversely proportional to the cross-sectional area. Increasing the length of the thermoelements may minimize its effect if L/A remains constant. However, economic considerations demand that both L and A be kept as small as possible. If there is significant contact resistance, reduction of the length beyond a certain value can so impair the performance of a module that the number of thermoelements has to be increased and more than offsets the saving of material in each element. Parrott and Penn showed how to maximize the cooling power for a thermocouple of given volume.

We suppose that the total electrical contact resistance of the couple is r_c for unit cross-sectional area. Then, (1.17) becomes

$$Q_C = (\alpha_p - \alpha_n)IT_C - K(T_H - T_C) - \frac{1}{2}I^2\left(R + \frac{r_c}{A}\right). \tag{5.15}$$

if A is regarded as the total cross-sectional area. The power consumption is

$$W = (\alpha_p - \alpha_n)I(T_H - T_C) + I^2\left(R + \frac{r_c}{A}\right). \tag{5.16}$$

Thus, we see that the effect in both equations is the same as if the distributed resistance of the branches were increased from R to $(R+r_c/A)$ Alternatively, if we do not differentiate between the resistivities of the p-type and n-type materials, we may regard the electrical resistivity as effectively increased from ρ to $(\rho+r_c/L)$.

There are two possible approaches to optimizing the device. The ratio of cooling power to volume can be found for either maximum cooling power or maximum coefficient of performance. This means that we can find the cooling power from (5.15), using the current for either Q_{max} or ϕ_{max} given in Table 1.2, there $R^* = (R+r_c/A)$ is substituted for R. The maximum cooling power per unit volume is given by

$$\frac{Q_{max}}{AL} = \frac{\lambda}{L^2}\left[\frac{RZT_C^2}{2R^*AL} - (T_H - T_C)\right], \tag{5.17}$$

where λ is assumed to be the same for both branches. If we assume that ΔT is considerably less than the absolute temperature at either end, (5.17) may be written as

$$\frac{Q_{max}}{AL} = \frac{2\lambda T}{r_c^2 \rho^2}\Gamma_q. \tag{5.18}$$

Here Γ_q is a function, defined by Parrott and Penn [5.109], that is given by

$$\Gamma_q = \frac{r_c^2}{\rho^2 L^2}\left[\frac{ZT}{4}\left(1 + \frac{r_c}{\rho L}\right)^{-1} - \frac{\Delta T}{2T}\right]. \tag{5.19}$$

The corresponding value for the coefficient of performance is

$$\phi_q = \left[ZT - 2\left(1 + \frac{r_c}{\rho L}\right)\frac{\Delta T}{T}\right]\left[2ZT\left(1 + \frac{\Delta T}{T}\right)\right]^{-1}. \tag{5.20}$$

The cooling power per unit volume at ϕ_{max} is

$$\frac{Q_\phi}{AL} = \frac{2\lambda T}{r_c^2 \rho^2} \Gamma_\phi, \qquad (5.21)$$

where the function Γ_ϕ is defined by

$$\Gamma_\phi = \frac{r_c^2 \Delta T}{2\rho^2 L^2 T} \cdots$$

$$\cdots \left\langle \frac{ZT}{1+\dfrac{r_c}{\rho L}} \left\{ \left[1 + ZT \dfrac{\left(1+\dfrac{\Delta T}{2T}\right)}{1+\dfrac{r_c}{\rho L}} \right]^{\frac{1}{2}} - 1 - \dfrac{\Delta T}{2T} \right\} \left\{ \left[1 + ZT \dfrac{\left(1+\dfrac{\Delta T}{2T}\right)}{1+\dfrac{r_c}{\rho L}} \right]^{\frac{1}{2}} - 1 \right\}^{-2} - 1 \right\rangle.$$

(5.22)

Then, the coefficient of performance is given by

$$\phi_{max} = \frac{T}{\Delta T} \left\{ \left[1 + ZT \dfrac{\left(\dfrac{1+\Delta T}{2T}\right)}{1+\dfrac{r_c}{\rho L}} \right]^{\frac{1}{2}} - 1 - \dfrac{\Delta T}{T} \right\} \left\{ \left[1 + ZT \dfrac{\left(1+\dfrac{\Delta T}{2T}\right)}{1+\dfrac{r_c}{\rho L}} \right]^{\frac{1}{2}} + 1 \right\}^{-1}.$$

(5.23)

To proceed further, we need specific values for ZT, $\Delta T/T$, and the dimensionless parameter $r_c/\lambda L$. By way of illustration, we shall apply the theory to a couple, that has ZT equal to unity and the relative temperature difference $\Delta T/T$ is equal to 0.1. We also consider the condition of maximum coefficient of performance, though qualitatively similar results are obtained if the condition of maximum cooling power is applied.

Figure 5.39 shows how the coefficient of performance falls as $r_c/\rho L$ rises when these values of ZT and ΔT are substituted in (5.22) and (5.23). The fact that ϕ_{max} becomes zero when $r_c/\rho L = 4$ indicates that a decrease in length below $r_c/4\rho$ makes it impossible to achieve a temperature difference as great as $0.1T$, even when no heat is extracted from the source. Of course, the volume of thermoelectric material for a given cooling power becomes infinite at this particular length.

Examination of Fig. 5.39 shows that there are generally two lengths that give the same value of Γ_ϕ and, thus, the same volume of thermoelectric material for a given cooling power. It is preferable to adopt the greater of these two lengths, because this

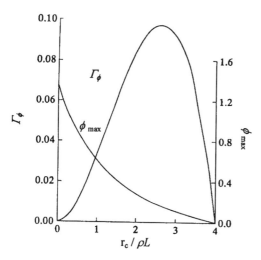

Fig. 5.39. Plots of the function Γ_ϕ and of the maximum coefficient of performance against $r_c/\rho L$ for a thermoelectric refrigerator with a significant electrical contact resistance. The plots are based on a thermocouple where $ZT = 1$ and the temperature difference ΔT is equal to $0.1T$

gives the higher ϕ_{max}. A value of $r_c/\rho L$ equal to 2.5 yields the greatest economy in the use of material. This is achieved at the cost of reducing the coefficient of performance to only about one-seventh of its value for zero contact resistance. In fact, the r_c of properly applied electrical contacts is very small. Even for thermoelements no more than 2 mm long, the value of $r_c/\rho L$ is expected to be less than 0.1.

We note that joints to most thermoelectric materials are made by applying solder to plated contacts. It is possible to make direct connections to bismuth telluride with a bismuth–tin solder, after first wetting the surface with bismuth. However, it is more satisfactory to nickel plate the ends of the thermoelements. The value of r_c for contacts plated to bismuth telluride is reportedly less than $10^{-9}\,\Omega\,\text{m}^{-2}$ [5.113].

5.5.4 Multistage Refrigerators

The maximum temperature difference that can be reached by a single-stage refrigerator is given in Table 1.2. It is, however, possible to obtain greater temperature differences by operating cooling units in cascade, that is, in a series–thermal arrangement. The first stage of the cascade provides a low-temperature heat sink for the second stage which, in turn, provides a sink at an even lower temperature for the third stage, and so on. Manufacturers have been able to reduce the thermal resistance between stages to such a low value that six-stage cooling units are available as standard production items. Typical arrangements for two-stage cascades are shown in Fig. 5.40. A characteristic feature of all cascades is their pyramidal shape, because it is invariably necessary that the cooling capacity of the higher temperature stages be much greater than that of those that operate at the lower temperatures. In an N-stage unit, the nth stage must have a cooling power that is equal to the sum of the cooling power at the source and the electrical power that is used in each of the stages N, $(N-1)$, $(N-2)$,…, and $(N-n)$.

Suppose that the coefficient of performance for the nth stage is ϕ_n. If Q_N is the rate of cooling for the Nth stage, the rate of cooling for the nth stage is $Q_N(1+1/\phi_N)(1+1/\phi_{N-1})\ldots(1+1/\phi_{N-n})$. The rate at which heat is delivered to the sink is $Q_N(1+1/\phi_N)(1+1/\phi_{N-1})\ldots(1+1/\phi_1)$, so that the overall coefficient of performance is

$$\phi = \left[\left(1+\frac{1}{\phi_N}\right)\left(1+\frac{1}{\phi_{N-1}}\right)\ldots\left[1+\frac{1}{\phi_1}\right]-1\right]^{-1}. \tag{5.24}$$

Although there is no reason why each stage should be operated with the same coefficient of performance, we shall assume that the cascade is designed so that this is the case. If the coefficient of performance of each stage is ϕ', then

$$\phi = \left[\left(1+\frac{1}{\phi'}\right)^N - 1\right]^{-1}. \tag{5.25}$$

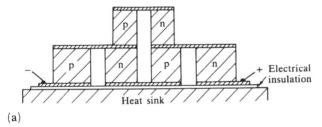

(a)

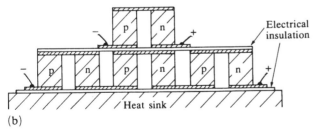

(b)

Fig. 5.40. Two types of two-stage thermoelectric cascades. In **(a)** the current divides itself between stages, whereas in **(b)** each stage consists of a separate module

It will be assumed that each stage operates at the maximum coefficient of performance, and ϕ is as given in Table 1.2.

We have applied (5.25) to determine the coefficient of performance of multistage refrigerators that have up to six stages. A conservative value for Z equal to 2.5×10^{-3} K^{-1} has been used in the calculations. This gives a minimum source temperature T_C of about 233 K for a single-stage module when the heat sink is at 300 K.

Assuming the same heat sink temperature of 300 K, the variation of ϕ with the heat source temperature for 1, 2, 3, 4, and 6-stage units is shown in Fig. 5.41. It is clear that there is a substantial gain in ϕ for source temperatures below about 250 K if a multistage unit is used. It is, of course, essential to use more than one stage when the source temperature is below 233 K, even when the required cooling power is negligible.

Also shown in Fig. 5.41 is the range of minimum values for the source temperatures obtained from the data sheets for multistage units manufactured by Marlow Industries, Inc. in the late 1980s. It appears that our theory can predict the minimum temperature that can be achieved in practice if we assume an overall coefficient of performance equal to about 0.01 to allow for unavoidable losses.

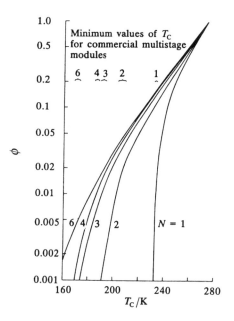

Fig. 5.41. Coefficient of performance plotted against source temperature for N-stage thermoelectric cascades, where $N = 1, 2, 3, 4,$ and 6. It is assumed that $Z = 2.5 \times 10^{-3}$ K^{-1} and that the sink is at 300 K. Also shown are the ranges of the source temperature for commercial modules produced by Marlow Industries, Inc.

The increased cooling power that is required for the high-temperature stages of a cascade is usually achieved by increasing the number of thermoelements rather than by decreasing their length–area ratio. For example, a six-stage cascade might consist of modules with 1, 2, 4, 8, 16, and 32 thermocouples.

As we have already shown, at low temperatures, bismuth–antimony alloys become superior to the alloys based on Bi_2Te_3. Therefore, it is often advantageous to incorporate negative bismuth–antimony thermoelements in the higher stages and to maintain these stages between the poles of a magnet. Yim and Amith [5.45] reported a cooling capacity of 8 mW, when the source was at 128 K and the sink at 300 K, from a hybrid unit that was based on the Peltier and magneto-Peltier effects in Bi_2Te_3 and bismuth–antimony alloys.

5.5.3 Thermomagnetic Refrigerators

Now, we turn to the design of thermomagnetic refrigerators. The problem differs from that of thermoelectric units because the directions of the flow of heat and electric current are perpendicular to one another. This means that, at least in principle, it is possible to achieve any required cooling power by using a single piece of thermomagnetic material, even if the applied current is limited.

First, let us consider a simple rectangular block of material, ignoring the effect of the end contacts on the equipotentials and isothermals. As pointed out earlier, the equations for thermoelectric cooling remain applicable with the substitutions NB_zL_x/L_y for $(\alpha_p - \alpha_n)$, $\delta L_xL_z/L_y$ for K, and $\Delta L_x/L_zL_y$ in place of R. It will be recalled that N is the Nernst coefficient, B_z is the magnetic field, and x and y indicate directions that are parallel to the current flow and temperature gradient, respectively. As before, ρ and λ are the adiabatic electrical resistivity and isothermal thermal conductivity, respectively, in the applied field.

The current I will be determined for the condition of maximum cooling power because this is more appropriate than that of the optimum coefficient of performance for this type of device. Then,

$$I_q = \frac{NB_zL_zT_c}{\rho}, \tag{5.26}$$

and the cooling power is

$$Q_{max} = \frac{\lambda L_xL_z}{L_y}\left(\frac{1}{2}z_ET_C^2 - T_H + T_C\right). \tag{5.27}$$

The potential difference across the sample is

$$V_q = \frac{NB_z L_x T_H}{L_y}. \quad (5.28)$$

Thermomagnetic cooling is likely to be useful only at low temperatures, so we consider a specific example in which the heat sink is maintained at 100 K. We give the data in Table 5.6 that are appropriate for a crystal of $Bi_{0.99}Sb_{0.01}$, as specified by Yim and Amith [5.45]. The maximum depression of temperature would be about 14 K, and it is supposed that the actual value of $(T_H - T_C)$ is 10 K. If we insert the values from Table 5.6 in (5.26) to (5.28), we find that $I_q=3.33 \times 10^3 L_z$ A, $Q_{max} = 57.35 L_x L_z / L_y$ W, and $V = 0.10 L_x / L_y$ V, where the lengths are expressed in meters.

Although it might seem that I_q and V can be adjusted independently to yield any required cooling power, there are certain practical considerations that must be borne in mind. For example, to minimize the effects of the end contacts, the length L_x in the direction of the electric current should be much larger than L_y in the direction of heat flow. It is also desirable that the gap between the pole pieces of the magnet should be short and of small cross-sectional area. Again, there is an upper limit to the size of the sample of thermomagnetic material that is set by the available crystals and a lower limit that is set by the difficulty of handling very small pieces. If L_z is equal to 1 mm, the current I_q has the reasonable value of 3.33 A. Also, if L_x / L_y is set at about 5, the cooling power, Q_{max}, is about 0.3 W, and the applied voltage is about 0.5 V. Such values may match the requirements for certain applications.

Next, we discuss an important feature of thermomagnetic refrigeration that stems from separating the flows of heat and electricity. Because cooling power is proportional to $L_x L_z / L_y$, it is possible to make a multistage unit of constant cross-section $L_x L_z$ normal to the heat flow by changing the value of L_y for each of the stages. Then, the cascade would take the form shown in Fig. 5.42(a). The current I can be the same for each stage, but the applied voltage must be greater for the stages that are

Table 5.6. Data used in designing a single-stage thermomagnetic refrigerator

Parameter[a]	Value
Z_E	4.0×10^{-3} K^{-1}
NB_z	1.0×10^{-3} V K^{-1}
λ	9.25 W m^{-1} K^{-1}
δ	2.7×10^{-5} Ω m
T_H	100 K

[a] These parameters have values that are close to those expected for the alloy $Bi_{0.99}Sb_{0.1}$ in a magnetic field of 0.75 T in the preferred orientation [5.45]

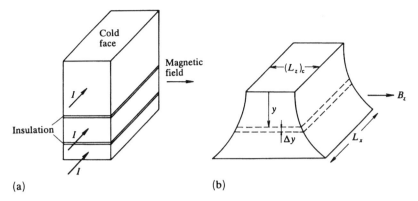

Fig. 5.42. Thermomagnetic cascades: (a) a three-stage cascade of uniform cross-section and (b) an infinite-staged cascade of exponential shape

nearer to the heat sink. This means, of course, that the stages must be electrically isolated from one another.

Such electrical isolation is unnecessary if the alternative approach, shown in Fig. 5.42(b), is adopted. Here, the width in the magnetic-field direction is allowed to vary. The cooling power can be increased, as one approaches the heat sink, by increasing L_z so that the potential difference V remains constant [5.114].

Consider a section of the device of thickness Δy that is displaced by y from the cold face, and suppose that the temperature difference across this section is ΔT. If we regard this section as one stage of the cascade, its optimum coefficient of performance is given by

$$(T/\Delta T)\left[(1+z_E T)^{1/2} - 1\right]\left[(1+z_E T)^{1/2} + 1\right]^{-1}. \tag{5.29}$$

Thus, the ratio of the heat rejected at $y + \Delta y$ to that taken in at y is

$$1 + (\Delta T/T)\left[(1+z_E T)^{1/2} + 1\right]\left[(1+z_E T)^{1/2} - 1\right]^{-1}. \tag{5.30}$$

This ratio must be equal to $(L_z + \Delta L_z)/L_z$. Thus

$$\frac{\Delta L_z}{L_z} = C_E \frac{\Delta T}{T}, \tag{5.31}$$

where

5. Review of Established Materials and Devices

$$C_E = \left[(1+z_E T)^{1/2} + 1\right]\left[(1+z_E T)^{1/2} - 1\right]^{-1}. \tag{5.32}$$

Setting up an integral equation on this basis, we obtain

$$\int \frac{dL_z}{L_z} = \int \frac{C_E}{T} dT = \int \frac{C_E}{T} \frac{dT}{dy} dy. \tag{5.33}$$

If it is assumed that $(C_E/T)(dT/dy)$ is approximately constant, (5.33) may be solved to yield

$$L_z = (L_z)_c \exp\left(\frac{C_E}{T} y \frac{dT}{dy}\right). \tag{5.34}$$

Thus, by making a sample of exponential shape, one can achieve a cascade with an infinite number of stages. A practical drawback is that the gap between the pole pieces of the magnet must be large enough to accommodate the thermomagnetic element at its widest point.

Horst and Williams [5.73] avoided the difficulty of making an exponentially shaped bar by using a simpler trapezoidal cross-section. The performance was downgraded by about 40% compared with that expected for the ideal exponential shape, but was still much better than that obtained from a rectangular bar. It was fabricated from $Bi_{0.97}Sb_{0.03}$, of very low excess-carrier concentration. In a magnetic field of 0.75 T, it yielded 75 mW of cooling at a temperature difference of about 35 K with the heat sink at 156 K.

6. The Phonon–Glass Electron-Crystal Approach to Thermoelectric Materials Research

6.1 Requirements for Good Thermoelectric Materials and the PGEC Approach

Based on the theory outlined in Part I, one can list the requirements for materials suitable for thermoelectric cooling and power generation applications at and above ambient temperatures. For semiconductors these are the basic criteria:
1. High crystal symmetry with the electronic bands near the Fermi level and many valleys.
2. Heavy element compounds with small electronegativity differences between constituent elements.
3. Energy gap of about 10 $k_B T$, where T is the temperature of operation and k_B is Boltzmann's constant.

Incorporating of these criteria into one material remains elusive. These criteria can be incorporated to obtain relatively good figures of merit; however, in most cases, the thermal conductivity remains relatively large. Several approaches and directions are currently underway in the search for new thermoelectric materials, some of which will be presented in the next three chapters. An approach that has drawn particular attention is one that attempts to "identify" semiconductor compounds that have good electronic properties and also very low thermal conductivity values.

The concept of a "phonon–glass electron-single-crystal" (PGEC), first introduced by Slack and given in detail in his review [6.1], is at the heart of the investigation into semiconductors that possess "glass-like" thermal conductivity values and may prove superior for thermoelectric applications. PGEC materials would possess electronic properties normally associated with good semiconductor single crystals but would have thermal properties normally associated with amorphous materials. The introduction of the PGEC concept to thermoelectrics is one of the more significant innovations in the field of thermoelectrics in the last 30 years. The utility of this approach is clear from the definition of z. In practice, however, such a material may be very difficult to "engineer" if not identify.

A key task in the PGEC approach is to have very low thermal conductivity solids. Cahill et al. [6.2] enumerated a number of crystalline systems that possess low, glass-like lattice thermal conductivity (λ_L) values. The thermal conductivities measured in these systems indicate that the lattice vibrations are similar to those of amorphous materials. The relationship between glass-like λ_L and the theoretical minimum thermal conductivity λ_{min} of any solid material was pointed out by Slack [6.3]. The main features of all of these materials and the main conclusion of the research by Cahill et al. [6.2] indicate that

1. They possess "loose" atoms or molecules whose translational or rotational positions are not well defined and possess two or more metastable positions.
2. There is no long-range correlation between the positions of the "loose" atoms or molecules.
3. The mass of these "loose" atoms or molecules is at least 3% of the total mass of the crystal.
4. Disorder produced by point defect scattering cannot lead to glass-like lattice vibrations; this approach will not lead to λ_{min}.

This approach has resulted in solids with very low thermal conductivities, but it is not clear that this same approach can be translated into semiconductors with good electronic transport properties. The type of "open structure" compound that may possess PGEC properties would contain at least three, perhaps more than three, distinct crystallographic sites so that there is a possibility of obtaining good electronic properties. At least one of the sites may reside inside an atomic "cage" formed by the constituent atoms at the other sites. The disorder caused by the "rattling" motion of this atom would scatter the heat-carrying phonons. The other sites would form the framework of the cage-like open structure. These framework atoms would presumably dominate the band structure that defines electronic transport properties. In this chapter, we review two material systems that have received enormous attention for thermoelectric applications. It is important to note that skutterudite and clathrate compounds have features in common with the crystalline systems outlined by Cahill et al. [6.2].

6.2 The Skutterudite Material System

6.2.1 Introduction

The name skutterudite derives from a naturally occurring mineral, $CoAs_3$, first found in Skutterud, Norway. This mineral possesses a cubic crystal structure and contains thirty-two atoms per unit cell. A schematic of the basic structure is shown in Fig. 6.1. Binary skutterudite compounds, such as $CoSb_3$, have interesting properties useful for thermoelectrics. Skutterudites possess a large unit cell, heavy constituent atom masses, low electronegativity differences between the constituent atoms, and large carrier mobilities. They form covalent structures with low coordination numbers for the constituent atoms and so can incorporate atoms in the relatively large voids formed. Therefore, compounds can be formed where atoms fill

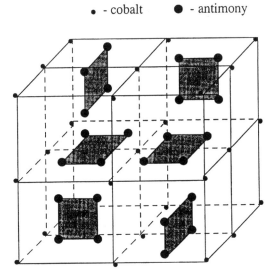

Fig. 6.1. Schematic of the skutterudite unit cell illustrating the two voids per unit cell and the nearly square pnicogen rings

the voids of the skutterudite structure. As first predicted by Slack and Tsoukala [6.4], interstitially placing atoms in the voids of such compounds would substantially reduce the thermal conductivity λ by introducing phonon-scattering centers. Thus, these atomic "void-fillers" may "rattle" about in their oversized cages and thereby provide an approach to drastically reducing λ in the high-λ binary compounds. Another advantage of this cubic material system is that single crystals are not necessary to investigate the electrical and thermal transport properties of skutterudites. This also makes their promise for thermoelectric applications more feasible, if the appropriate material parameters are achieved. In addition, these materials are hard and have a relatively low coefficient of thermal expansion. These factors demonstrate why these materials continue to be of interest for thermoelectric applications.

6.2.2 Approach to Synthesis

Binary antimonides form peritectically [6.5]. Therefore, these compounds can be readily synthesized by using standard powder metallurgical approaches. Single crystals have also been grown from an antimony-rich melt [6.6]. Filled skutterudites have been most readily synthesized in polycrystalline form. One of the new and more interesting synthetic approaches recently developed has led to the formation of new skutterudite compounds.

Modulated elemental reactants, a new synthetic approach pioneered by D.C. Johnson at the University of Oregon, permits synthesizing compounds that are metastable relative to disproportionation [6.7]. The starting multilayer reactants are made by sequential vapor phase deposition of the elements in a high vacuum system. The key advantage of using multilayer elemental reactants is the ability to control diffusion distances through the design of the initial reactants. If the individual layer thicknesses are less than a critical thickness, the multilayer reactants interdiffuse to form a bulk amorphous alloy. Subsequent annealing results in the crystallization of the easiest compound to nucleate. The composition of the amorphous intermediate provides a means of controlling relative nucleation energies to favor the formation of a desired compound. The amorphous state is similar to a melt but has the added advantage of being diffusionally constrained.

The ability to avoid thermodynamically more stable compounds by controlling reaction intermediates is exceedingly important in preparing new compounds. If one can avoid the formation of thermodynamically stable binary compounds as reaction intermediates, the number of new compounds that can be prepared increases dramatically. In the case of skutterudite compounds, this has been clearly demonstrated by the preparation of thermodynamically unstable binary compounds such as $FeSb_3$ [6.8]. The ability to avoid thermodynamically stable binary compounds increases in ternary and higher order systems. The ability to avoid binary compounds is crucial in preparing ternary compounds that are thermodynamically unstable with respect to the binaries. The ternary iron–antimonides are examples of the synthesis of such kinetically stable compounds using multilayer reactants [6.8–6.10].

Although most studies using elementally modulated reactants have investigated fundamental questions concerning the way they evolve to crystalline products, the demonstrated ability of this approach to prepare thermodynamically unstable compounds via a relatively straightforward synthetic route is of practical importance for other synthetic approaches. For example, if one has access only to thermodynamically stable compounds, one does not need combinatorial libraries to determine the stable compounds in a phase diagram. However, most likely, many compounds in higher order phase diagrams can be accessed only if one can avoid known binary compounds. Elementally modulated reactants provide a route to these compounds, where the diffusionally constrained nature of the amorphous state prevents disproportionation to the thermodynamically more stable products. It is easy to make combinatorial libraries of layered starting reactants by using simple shutter systems. Because the composition of the amorphous intermediate determines nucleation kinetics, a combinatorial approach involving parallel processing of many samples is an attractive way to explore compositional space. This approach may lead to new materials, perhaps of interest for thermoelectric applications.

6.2.3 Crystal Structure

The basic family of binary semiconducting compounds that form the skutterudite structure can be represented by MX_3 where M = Co, Rh, or Ir and X = P, As, or Sb. There are eight formula units per crystallographic cubic unit cell in the space group $Im3$, eight M atoms occupy the c crystallographic sites, and twenty-four X atoms are situated at the g sites. The structure can be uniquely specified from the lattice parameter, and the two positional parameters y and z specify the g site. One of the salient features that characterizes the skutterudite structure is the nearly square X_4 rings that run parallel to the cubic crystallographic axes. From structural analysis of binary skutterudites, the void radii of these compounds, it has been estimated, range from 1.763 to 2.040 Å for CoP_3 and $IrSb_3$, respectively [6.11].

As mentioned before, a prominent feature of the skutterudite material system is the fact that it has an "open structure" lattice; two relatively large voids exist at the a positions of the unit cell which can be interstitially filled with atoms. The cubic unit cell can then be written in a general way as $\square_2M_8X_{24}$, where $\square$ represents a void. Filled skutterudites have been formed where group-III, group-IV, lanthanide, actinide and alkaline-earth ions interstitially occupy these voids [6.12]. The interstitial ion in this structure is sixfold coordinated by the X-atom planar groups and is thereby enclosed in an irregular dodecahedral (12fold coordinated) "cage" of X atoms. An illustration of the unit cell of a filled skutterudite centered at the position of one of the interstitial "guest" ions is shown in Fig. 6.2. This configuration of filled skutterudites is one of the most conspicuous aspects of the structure and directly determines many of their physical properties, as will be discussed later.

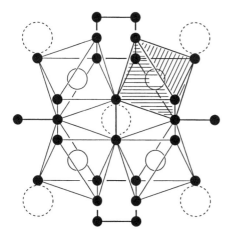

Fig. 6.2. The skutterudite unit cell centered at the void-filler atom, which is enclosed in an irregular dodecahedral (12fold coordinated) cage of pnicogen (filled circles) atoms

Large X-ray atomic displacement parameters (ADP) have been reported for the "guest" ions, indicating that they may "rattle" or participate in soft phonon modes in the voids of this crystal structure [6.13–6.15]. The ADPs of these ions increase with decreasing ionic size. It appears these ions are "caged" in the voids of this structure and, if smaller than the void in which they are caged, may "rattle" and thus interact randomly with the lattice phonons resulting in substantial phonon scattering. This is most evident from temperature-dependent structural data. The ADP data for La, Fe, Co, and Sb obtained from single-crystal neutron diffraction of $La_{0.75}Fe_3CoSb_{12}$ are shown in Fig. 6.3 [6.16]. The ADP values for Fe, Co, and Sb are typical for these elements in the skutterudite compounds and indicate that the (Fe,Co)Sb$_6$ framework is essentially rigid. The ADP values for La are large and have strong temperature dependence. This indicates that La is "rattling" about its equilibrium position with an average amplitude of 0.015 nm at room temperature, thereby creating localized dynamic disorder. This dynamic disorder is the cause of the unique thermal transport in these compounds.

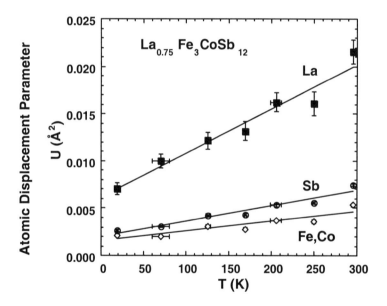

Fig. 6.3. Temperature dependence of the isotropic displacement parameters measured on a single crystal of $La_{0.75}Fe_{2.74}Co_{1.26}Sb_{12}$. Reprinted from [6.15]. Copyright 1997, American Physical Society

6.2.4 Bonding and Band Structure

Given the structural information, a simple bonding arrangement can be used to describe this structure [6.17– 6.19]. Each X atom bonds with its two nearest neighbors in the X_4 groups via σ bonds, thereby involving two of its valence electrons. The remaining valence electrons of the X atoms (three per atom, five total) participate in the two M–X bonds. Because each metal atom is octahedrally coordinated by X atoms, a total of (3/2)6 = 9 X electrons is available for bonding in each MX_6 octahedron. The M atoms (having a s^2d^7 configuration) can provide an additional nine electrons for a total of eighteen electrons in this arrangement. Of the eighteen electrons available for bonding, six fill the nonbonding orbitals in a spin-paired arrangement, and the remaining twelve fill bonding orbitals. These compounds possess no unpaired spins and therefore should be diamagnetic semiconductors, as has been reported. In NiP_3, this is not the case and leads to metallic conductivity and paramagnetism [6.18].

In filled skutterudites, electron-deficient elements can be substituted at the metal or pnicogen sites when the void is filled in the skutterudite structure by lanthanide, actinide, or alkaline-earth atoms. For example, in $LaFe_4P_{12}$ for iron in the low spin Fe^{2+} (d^6) state, the formula can be written as $La^{3+}[Fe_4P_{12}]^{4-}$. With the transfer of three electrons from La^{3+}, the polyanion is just one electron deficient and thus $LaFe_4P_{12}$ should be metallic. For $CeFe_4P_{12}$, if it is assumed that the cerium ion is in the tetravalent state, Ce^{4+}, one has $Ce^{4+}[Fe_4P_{12}]^{4-}$ such that the $CeFe_4P_{12}$ polyanion is isoelectronic with CoP_3 and thus $CeFe_4P_{12}$ should be semiconducting. For the case of pnicogen substitution, one may substitute a group-IV element on the pnicogen site, for example, $La^{3+}[Co_4Sn_3Sb_9]^{3-}$.

Although these simple arguments are useful, band-structure calculations may provide a more realistic physical picture [6.12]. In general, these calculations essentially agree with simple predictions of the two-electron-bond model described before. The binary semiconducting skutterudites possess zero or very narrow band gaps, and the conduction band is very flat, implying a large electron effective mass and the potential for excellent thermoelectric properties in n-type materials. Electronic band-structure calculations have also been carried out on filled skuterudties. Calculations of Singh and Mazin [6.20] indicated that that the band structure of $LaFe_3CoSb_{12}$ displays an indirect gap of approximately 0.6 eV. Similar to the conclusions of Harima [6.21], the highest occupied state is derived from bands with mostly Sb p character, but there are Fe/Co derived d bands, whose tops are situated only about 100 meV below the tops of the p-like states. These d band states have very low dispersion and therefore high effective mass which gives rise to the large observed Seebeck coefficients. Pnicogen site substitution is not as straightforward because these bonds directly affect the band structure.

6.2.5 Thermal Conductivity

Although it had been known for some time that $CoSb_3$ has a relatively large λ, partly due to its highly covalent cubic crystal structure, investigation into the thermal properties of compounds that have the skutterudite crystal structure was begun in earnest because this material system has potential for thermoelectric applications [6.12]. For example, crystalline $CoSb_3$ samples display behavior typical of simple solids whose magnitude is similar to that of InSb. Similarly, Slack and Tsoukala [6.4] showed that the compound $IrSb_3$ and the alloy $Ir_{0.5}Rh_{0.5}Sb_3$ displayed a typical crystalline temperature dependence with the lattice component of λ, λ_L, of approximately 16 W m^{-1} K^{-1} for $IrSb_3$ at room temperature and 9.1 W m^{-1} K^{-1} for $Ir_{0.5}Rh_{0.5}Sb_3$ [6.4]. These values are much higher than that calculated for λ_{min} of $IrSb_3$. This led the authors to postulate that filled skutterudites may result in a large reduction in λ_L due to strong scattering from the "rattling" of the void-filling atoms in their oversized local sites in the crystal structure [6.4].

The major effort in the investigation of the skutterudite crystal system, particularly for thermoelectric applications, is due to the fact that atoms placed in the voids of this structure have a large effect on phonon propagation. Nolas et al. [6.11] investigated the effect on λ_L of a series of lanthanide-filled skutterudites. In Fig. 6.4, λ_L versus temperature is plotted for three lanthanide-filled polycrystalline skutterudites [6.11]. Also shown in the figure is the calculated λ_{min} of $IrSb_3$. As shown in Fig. 6.4, there is an order-of-magnitude decrease in λ_L at room temperature, compared to $IrSb_3$, and an even larger decrease at lower temperatures. In addition, the smallest most massive lanthanide ion results in the lowest λ_L. These smaller ions are freer to "rattle" inside the voids of the skutterudite structure, compared to La^{3+}, and thereby can interact with lower frequency phonons than the La^{3+} ions. In addition, the smaller lanthanide ions may introduce a static disorder in the lattice at lower temperatures due to the fact that these ions are more off-center than the larger ions. This "guest ion"–phonon coupling is an effective phonon-scattering mechanism. In addition, the low-lying $4f$ electronic levels Nd^{3+} and Sm^{3+} also produce additional phonon scattering, further reducing λ_L [6.11]. This effect is most prominent in the Nd-filled skutterudite of the three samples shown in Fig. 6.4 due to the low-lying ground-state energy levels of Nd^{3+}. Although the f-shell transitions are relatively weak phonon scatterers, compared to the disorder introduced by the lanthanide ion in the void of this structure, the effect is quite clear, as shown in Fig. 6.4, particularly at lower temperatures.

Much of the work on the thermoelectric properties of filled skutterudites has been devoted to the $RFe_{4-x}Co_xSb_{12}$ series of compounds, where R represents a lanthanide ion. Figure 6.5 shows λ_L versus temperature for polycrystalline $CoSb_3$, $CeFe_3CoSb_{12}$, and $LaFe_3CoSb_{12}$ from Sales et al. [6.15]. The electronic component of λ, λ_e, of the filled skutterudites shown in Fig. 6.5 was rather large,

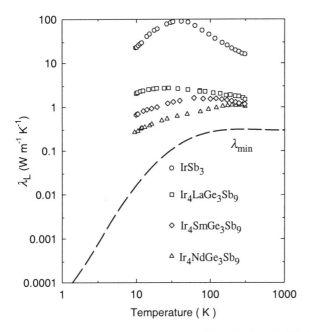

Fig. 6.4. Lattice thermal conductivity of three lanthanide-filled skutterudites, IrSb$_3$, and λ_{min} of IrSb$_3$. Reprinted from [6.11]. Copyright 1996, American Institute of Physics

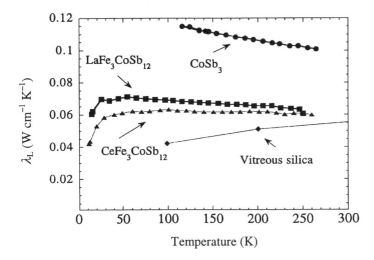

Fig. 6.5. Lattice thermal conductivity of Ce and La-filled skutterudites, CoSb$_3$, and amorphous SiO$_2$ (vitreous silica). Reprinted from [6.15]. Copyright 1997, American Physical Society

approximately 40% of λ, confirming that these compounds are very heavily doped. As discussed before, these materials possess low λ_L values due to the dynamic "rattle" disorder induced by the R atoms.

Nolas et al. [6.22] showed that a small concentration of La or Ce in the voids of $CoSb_3$ results in a relatively large reduction in λ_L. This can be easily seen from Fig. 6.6 which shows $La_xCo_4Sb_{12}$ with concentrations x = 0, 0.5, and 0.23. Room temperature values for $Ce_{0.1}Co_4Sb_{12}$ [6.23] and $Tl_{0.22}Co_4Sb_{12}$ [6.24] are also shown in the figure. Very little La, Ce, or Tl is required to reduce λ_L substantially. In the case of Ce-filled Fe-containing compounds, Meisner et al. [6.25] reported that these compounds can be thought of as solid solutions of $CeFe_4Sb_{12}$ and $\Box Co_4Sb_{12}$. All of these results indicate that partial void filling may be employed to reduce λ_L substantially. In fact, this demonstrates an interesting approach for further optimizing the transport properties for thermoelectric applications, as we will discuss later.

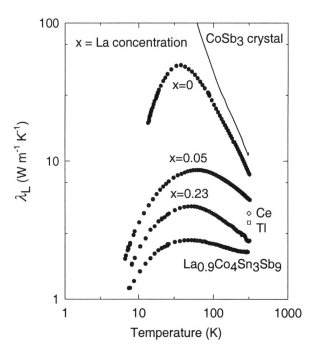

Fig. 6.6. Lattice thermal conductivity of $La_xCo_4Sb_{12}$ with x = 0, 0.05, and 0.23, single crystal $CoSb_3$, $La_{0.9}Co_4Sn_3Sb_9$, $Ce_{0.1}Co_4Sb_{12}$ (labeled Ce), and $Tl_{0.22}Co_4Sb_{12}$ (labeled Tl)

A rather interesting phonon-scattering mechanism that has been observed in skutterudites is that attributed to the multivalency of a constituent that forms the compound. The novel feature of this effect is that it appears to be due to the charge transfer of a pair of electrons of opposite spin. Nolas et al. [6.26] employed near-edge extended absorption fine structure analysis to determine the valence states of the constituents in $Ru_{0.5}Pd_{0.5}Sb_3$. It was observed that ruthenium in this compound is in 4+ and 2+ valence states in approximately a 50–50 ratio. The model, which describes the low λ_L measured in $Ru_{0.5}Pd_{0.5}Sb_3$ (λ_L = 2.9 W m^{-1} K^{-1} compared to 10.5 W m^{-1} K^{-1} for $CoSb_3$), as well as that observed for other Ru-containing skutterudite compounds [6.27], is that of phonon scattering due to the transfer of an electron pair in the d shell of Ru^{2+} to a neighboring Ru^{4+} [6.2]. Therefore, the scattering depends on the presence of both charge states. This is different than that due to spin disorder observed in some transition-metal fluorides and oxides.

6.2.6 Electronic Properties

The electronic transport properties of binary skutterudites were first studied in detail by Dudkin and co-workers in the Soviet Union in the late 1950s and early 1960s and centered on the properties of $CoSb_3$ [6.12]. An energy gap of 0.5 eV was determined from temperature-dependent conductivity measurements [6.28]. This result is similar to the indirect gap estimated some 40 years later by the band-structure calculations described before [6.29]. The electronic properties of binary skutterudites began to receive close scrutiny as potential thermoelectric materials when it was shown that these compounds can support high hole mobilities. Caillat et al. [6.30] reported single crystals of $RhSb_3$ with hole mobility as high as 10^4 cm^2 V^{-1} s^{-1} and $CoSb_3$ with mobilities approaching 2×10^3 cm^2 V^{-1} s^{-1}. All compounds possess large Seebeck coefficients consistent with semiconducting behavior. Sharp et al. [6.31] studied $Co_{1-x}Ir_xSb_{3-y}As_y$ alloys doped n-type with Ni, Te, or Pd and p-type with Fe, Ru, Os, and Ge and reported relatively high power factors $\alpha^2\sigma$, where α is the Seebeck coefficients and σ is the electrical conductivity. Samples doped n-type exhibited negative Seebeck coefficients at low temperature that crossed over to positive values above 500 K, indicating the onset of intrinsic conduction and the predominance of the higher mobility holes over the electrons.

Because of the very desirable heat conduction properties of filled skutterudite antimonides for thermoelectric applications, there was strong motivation to optimize the power factor of filled skutterudites. Thus Sales et al. [6.15, 6.32] and Chen et al. partially replaced Fe with Co and did indeed observe a return to a semiconductor-like resistivity and an increase in the Seebeck coefficient. Morelli et al. [6.23] and later Sales et al. [6.15] noted that replacing Fe by Co in $CeFe_4Sb_{12}$ reduces the amount of Ce filling the voids and that Fe/Co alloying and Ce filling are interrelated. They were able to achieve n-type doping at low Fe concentrations.

Note that even filled skutterudites that possess "metallic" resistivity and large (10^{21} cm^{-3}) hole density tend to have large α, implying hole effective masses on the order of the free electron mass, whereas n-type filled skutterudites have effective electron masses of several free electron masses. These observations are illustrated in Fig. 6.7 which shows high temperature transport properties of CeFe$_{4-x}$Co$_x$Sb$_{12}$ alloys as the Fe/Co ratio x is varied [6.32].

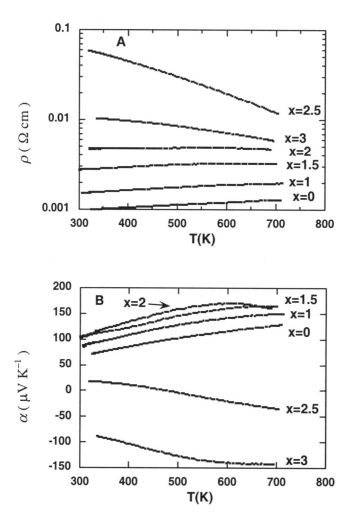

Fig. 6.7. Resistivity (**A**) and Seebeck coefficient (**B**) measurements of CeFe$_{4-x}$Co$_x$Sb$_{12}$ above room temperature. Reprinted from [6.32] with permission. Copyright 1996, American Association for the Advancement of Science

One approach that has reportedly been an optimization route for n-type materials is partial void filling [6.22–6.24]. The goal in this research is to obtain compounds with low thermal conductivity while maximizing the electronic properties. Nolas et al. [6.22] incorporated up to 23% La in $CoSb_3$ without metal or pnicogen site substitution, and Sales et al. [6.24] incorporated 22 % Tl. These partially filled skutterudites obey a rigid-band model where the void fillers act as electron donors. In n-type compounds, the electron effective mass is approximately an order of magnitude larger than that of the holes in p-type compounds. Thus, these compounds are extraordinary as thermoelectric materials in that they achieve optimum properties in a carrier concentration range much more reminiscent of a metal or semimetal than a moderately doped semiconductor. As indicated by Nolas et al. [6.22], these results indicate that partial filling of the voids is an interesting approach to optimizing the electronic properties of n-type compounds.

Substitutions at the pnicogen site were also employed to charge compensate for the trivalent lanthanide ions [6.17, 6.22, 6.24]. However, this approach resulted in metallic electronic conduction, perhaps due to the perturbed band structure; the Sb bonds were integral in forming the band structure in these compounds. This approach may be more successful when small concentrations of the substituted elements are used to optimize partially filled skutterudites.

6.2.7 Thermoelectric Properties

The concept of introducing "guest" atoms into the voids of the "open structure" of these materials essentially to "rattle around" and effectively scatter phonons has resulted in greatly reducing their thermal conductivity. This has resulted in relatively high zT values for these skutterudite materials at elevated temperatures. In Fig. 6.8, a plot of zT as a function of temperature for two optimized p-type polycrystalline filled skutterudites, from Sales et al. [6.15], is shown. The model calculation shown in Fig. 6.8 assumes parabolic bands, and the scattering rate r is predominated by acoustic phonons ($r = -1/2$). As seen from this figure, $zT = 1$ is reached at ~ 700K, and using the model, a maximum $zT = 1.4$ is reached at 1000 K.

As mentioned before, partial void filling is another approach that may lead to higher figures of merit, particularly in n-type skutterudites. Figure 6.9 shows zT as a function of temperature for polycrystalline $Yb_{0.19}Co_4Sb_{12}$. It has been previously reported that Yb in $YbFe_4Sb_{12}$ possesses an intermediate valence [6.33, 6.34]. Magnetic susceptibility measurements on $Yb_{0.19}Co_4Sb_{12}$ also indicate that this is the case with Yb which is intermediate between 2+ and 3+ [6.35]. This is advantageous for thermoelectric applications because more Yb can be interstitially placed in the voids of the skutterudite structure, compared to strictly trivalent lanthanide ions, while maintaining a similar carrier concentration, that is, a larger filling fraction in Yb-filled skutterudites compared to Ce-filled skutterudites, which

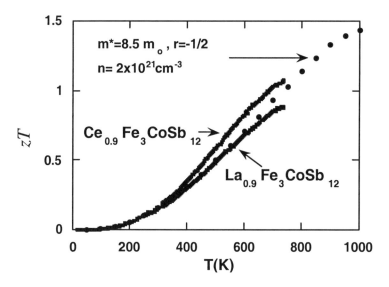

Fig. 6.8. zT versus temperature for Ce- and La-filled skutterudites. The experimental data are extrapolated to higher temperatures using a theoretical model. Reprinted from [6.32] with permission. Copyright 1996, American Association for the Advancement of Science

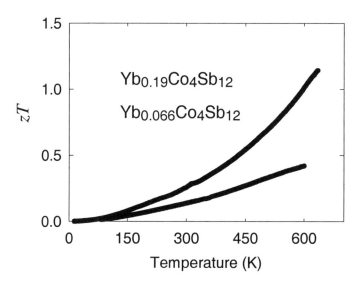

Fig. 6.9. zT versus temperature for $Yb_{0.19}Co_4Sb_{12}$ (upper curve) and $Yb_{0.066}Co_4Sb_{12}$ (lower curve). Above room temperature, a theoretical model was employed to extrapolate the lattice thermal conductivity along with experimental Seebeck and resistivity data

are trivalent above cryogenic temperatures, will result in a similar power factor while possessing substantially lower λ_L, thus further optimizing the dimensionless zT. As shown in Fig. 6.9, $zT \sim 0.3$ at room temperature, and $zT \sim 1$ above 600 K [6.35]. This compound has one of the highest zT values known to date in this temperature range.

The skutterudites are a fascinating family of materials that exhibit a wide range of electrical and thermal transport phenomena. In addition to obtaining thermoelectric materials with some of the best zT values above room temperature, the research into filled skutterudite compounds has also increased understanding of thermal transport phenomena and revealised novel phonon-scattering mechanisms. The success with skutterudite materials has facilitated the investigation of other new and novel systems with potential for thermoelectric applications. The "rattling" of the "guest" atoms in the voids of the skutterudites has prompted many researchers to seek materials with similar "open structures" in an attempt to obtain low λ_L while maintaining high power factors. One of the most promising of these classes of material is the semiconducting clathrates.

6.3 Clathrate Compounds

6.3.1 Introduction

That the crystalline complexes of water (H_2O) form clathrate compounds with simple molecules such as chlorine (Cl_2), has been known to for more than a century [6.36]. Many different clathrate hydrates have been discovered in nature and have been synthesized in the laboratory. These ice compounds have been formed with various materials such as noble gases, halogen molecules, some hydrocarbons, and low molecular weight compounds such as sulfur dioxide, methyl iodide, and chloroform, to name a few. In these clathrate hydrates, or ice clathrates, the water molecules form a hydrogen-bonded framework where each water molecule is tetrahedrally bonded to four H_2O neighbors, such as in normal ice, but with a more open structure that forms different types of cavities that can enclose atoms or molecules.

In the 1960s, Cros and co-workers [6.37] reported the existence of two clathrate phases, Na_8Si_{46} and $Na_{24}Si_{136}$, isomorphic with that of clathrate hydrates. Silicon, germanium, and later tin [6.38] form clathrate structures in which the guest atoms are alkaline atoms. In these materials, the host lattice is formed by atoms of one kind bonded by strong covalent forces with bond lengths similar to those in diamond-structured Si, Ge, or Sn.

The two most common forms of clathrates are known as type I and type II. The type I structure has a cubic cell and holds forty-six group-IV atoms (i.e., Si, Ge, or

Sn) in the unit cell. There are two types of cavities in the unit cell; two smaller pentagonal dodecahedra, where twenty water molecules are arranged to form twelve pentagonal faces, and six tetrakaidecahedra, where twenty four atoms are arranged by forming twelve pentagons and two hexagons. There are eight cavities, or atomic "cages" (M), in total in the cubic unit cell, and the general formula is 8M-46(IV). The type II clathrate also has a cubic cell, and there are 136 atoms per unit cell. These atoms are arranged in sixteen pentagonal dodecahedra and eight hexakaidecahedra, where the cage walls are twelve pentagons and four hexagons are formed from twenty-eight water molecules. There are twenty-four cavities in total per unit cell, and the general formula is 24M-136(IV). Figure 6.10 illustrates the polyhedral "building blocks" that form type I and II clathrates.

One of the more interesting discoveries in thermoelectrics was the magnitude and temperature dependence of the λ_L of Ge clathrates, as shown by Nolas et al. [6.39]. Figure 6.11 shows λ as a function of temperature from 0.6 to 300 K for a typical phase-pure polycrystalline $Sr_8Ga_{16}Ge_{30}$ specimen with large average grain size and high resistivity (i.e., assuming the Wiedemann–Franz relationship, essentially $\lambda \cong \lambda_L$ for this sample). Also shown in Fig. 6.11 is λ for single-crystal Ge, amorphous Ge (a-Ge), amorphous SiO_2 (a-SiO_2), and λ_{min} calculated for Ge [6.39]. We note that λ is lower than that of a-SiO_2 above 100 K, it is close to that of a-Ge at room temperature, and it exhibits temperature dependence that is reminiscent of amorphous materials. The low-temperature (<1 K) data indicate a T^2 temperature dependence, as shown by the straight line fit to the data in Fig. 6.11. Higher temperature data show a minimum, or dip, in the 4 to 35 K range indicative of possible resonance scattering. The low-temperature data also have a T^2 dependence on temperature similar to that found for amorphous material and similar to the T^2 dependence calculated for λ_{min} [6.3, 6.40]. It is clear from these data that the traditional alloy phonon scattering in the $Sr_8Ga_{16}Ge_{30}$ compound, which predominantly scatters the highest frequency phonons, has been replaced by one or more much lower frequency scattering mechanisms. The highest frequency optic phonons in the clathrate structure have very low or zero group velocity and contribute little to the total thermal conductivity, whereas the low-frequency acoustic phonons have the highest group velocity and contribute most to λ. The scattering of these low-frequency acoustic phonons by the encaged ions results in low thermal conductivities.

As will be discussed in detail later, the fact that clathrate compounds can possess "glass-like" thermal conductivity, have the ability to vary the electronic properties with doping level, and the relatively good electronic properties obtained in these semiconductor materials indicate why this material system continues to be of interest for thermoelectric applications. Besides the possibility of thermoelectric applications, these clathrate systems have many other interesting properties that

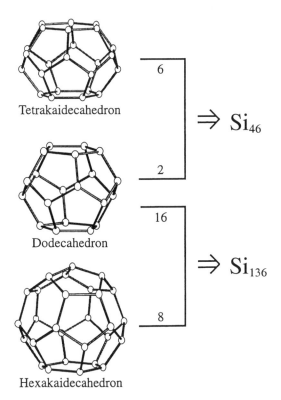

Fig. 6.10. The polyhedral "building blocks" of type I and II clathrates. The type I structure is formed by two pentagonal dodecahedra (top) and six lower symmetry tetrakaidecahedra (middle) in the cubic unit cell connected by shared faces. The type II structure is formed by sixteen dodecahedra (top) and eight hexakaidecahedra (bottom) also connected by shared faces in its larger cubic unit cell. Reprinted from [6.47], with permission

might lead to an entirely different range of applications from superconductivity [6.41, 6.42] to large band-gap semiconductors [6.43]. We note that semiconductor clathrates have been formed with crystal structures other than those of type I and II. For example $Ba_8Ga_{16}Sn_{30}$ forms a cubic, cage-like structure with symmetry $I43m$ and a lattice parameter of 11.5945 Å [6.44, 6.45]. Nolas et al. [6.46] showed that this compound is a semiconductor with low thermal conductivity.

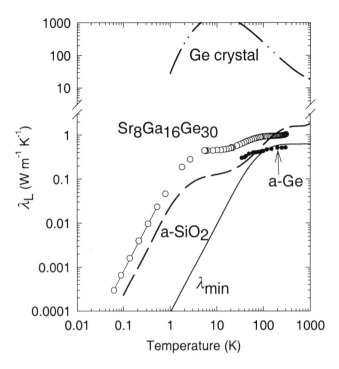

Fig. 6.11. Lattice thermal conductivity vs. temperature for polycrystalline $Sr_8Ga_{16}Ge_{30}$, single crystal Ge, amorphous SiO_2, amorphous Ge, and λ_{min} for Ge. The straight line fit to the $Sr_8Ga_{16}Ge_{30}$ data below 1 K produces a T^2 temperature dependence characteristic of glasses. Reprinted from [6.39] (© 1998 IEEE)

6.3.2 Approaches to Synthesis

There are several methods employed in forming Si, Ge, and Sn clathrates, as well as mixed-crystal clathrates that contain two or more of these elements. When Cros [6.37] first prepared silicon and germanium clathrates, he did so by thermal decomposition of the precursor MeSi or MeGe Zintl phases, where Me represents alkali metal ions. Later Ramachandran et al. [6.47] followed a similar method. To obtain clathrates, NaSi is first synthesized, then placed on a Ta boat and heated to 375°C under a 10^{-6} torr vacuum. Then, part of the sodium evaporates; thus the clathrate composition is controlled by varying the heating time from 30 minutes to several hours. The fractional filling of the cages diminished for longer heating times. Reny et al. [6.48] followed approximately the same procedure by heating under vacuum at temperatures between 340°C and 440°C to get type-II clathrates with less than fourteen sodium atoms per formula unit. For higher sodium concentrations, they used a closed steel reactor where they annealed Na_xSi_{136} (x<14) and excess Na in a temperature range from 370°C to 440°C. In these synthetic approaches, the reaction product was sometimes a mixture of the two different clathrate structures: the stoichiometric type-I (Na_8Si_{46}) and the variable sodium content type-II Na_xSi_{136}. Washing with ethanol and water then reacts with and removes all of the undecomposed NaSi. The clathrate itself does not decompose in water because the sodium is protected by its silicon cage. Ramachandran et al. [6.47] obtained pure phases of the type II clathrate (that is, only grains with a certain preset Na filling) by density separation of powder samples in a solution of dibromomethane (with a density of 2.477 g cm^{-3}) and tetrachloroethylene (1.614 g cm^{-3}) . The density of Na_xSi_{136} varies between 2.05 and 2.33 g cm^{-3} and depends on the Na concentration x.

It is also possible to produce clathrate compounds based on silicon, germanium, or tin networks by more standard techniques. Eisenmann et al. [6.44] prepared ternary type-I compounds with composition $X_8Y_{16}Z_{30}$, (where X is Sr or Ba, Y is Al or Ga, and Z is Si, Ge, or Sn in this formula) by mixing stoichiometric quantities of the pure elements in alumina crucibles at appropriate temperatures. Nolas et al. [6.49] grew single crystals of $Sr_8Ga_{16}Ge_{30}$ and $Eu_8Ga_{16}Ge_{30}$ by mixing and reacting stoichiometric quantities of the high purity elements. $Sr_8Ga_{16}Ge_{30}$ melts congruently at 760°C [6.50]. The resulting aggregate consisted of single-crystal grains of the order of 1–3 mm long which were stable in air and water. Chakoumakos et al. [6.51] grew single crystals of $Sr_8Ga_{16}Ge_{30}$ by first arc melting high-purity Sr and Ge together in an argon atmosphere to form $SrGe_2$, then reacting stoichiometric amounts of $SrGe_2$ with elemental Ga and Ge in a carbonized silica tube. Single crystals of $Sr_8Ga_{16}Ge_{30}$ 5–10 mm long were obtained.

Single crystals of halogen-filled clathrates were grown by Chu et al. [6.52] by vapor transport in a two-zone furnace. Nolas et al. [6.53, 6.54] grew small crystals of different Sn clathrates by mixing and reacting the constituent elements for two weeks at 550°C inside a tungsten crucible which was itself sealed inside a stainless steel canister after the canister was evacuated and backfilled with high-purity argon. The resulting Sn clathrates consisted of small octahedrally shaped crystals with a shiny, somewhat blackish, metallic luster. These millimeter-sized crystals were generally not reactive in air or moisture.

6.3.3 Crystal Structure, Bonding, and Band Structure

Clathrate compounds form in a variety of different structural types; however, the majority of work thus far on clathrates for thermoelectric research has been on two structural types that are isotypic with the clathrate hydrate crystal structures of type I and II [6.36]. The type I structure can be represented by the general formula X_8E_{46} and that of the type II structure by the formula $X_8Y_{16}E_{136}$, where X and Y are typically alkali-metal or alkaline-earth atoms. E represents a group-IV element Si, Ge, or Sn, although Zn, Cd, Al, Ga, In, As, Sb, or Bi can also be substituted to some degree for these elements. The key characteristic of both structural types is that the framework is formed by covalent, tetrahedrally bonded E atoms comprised of two different face-sharing polyhedra. As first pointed out by Slack [6.1], if atoms that are trapped inside these polyhedra, or atomic "cages", are smaller than these "cages", they may "rattle" about and interact randomly with the lattice phonons, resulting in substantial phonon scattering. This is one of the most conspicuous aspects of these compounds and directly determines many of their interesting and unique properties, including their thermoelectric properties, as will be described in detail later. As outlined in Section 6.3.1, these compounds display an exceedingly rich number of physical properties that result directly from the nature of their structure and bonding. The type I structure (Fig. 6.12) is comprised of two pentagonal dodecahedra, E_{20}, creating a center with $\bar{3}m$ symmetry, and six tetrakaidecahedra formed by twelve pentagonal and two hexagonal faces, E_{24}, creating a center with $\bar{4}m2$ symmetry. The corresponding unit cell is cubic with space group $Pm\bar{3}n$. The type II structure (Fig. 6.13) is comprised of sixteen pentagonal dodecahedra and eight hexakaidecahedra formed by twelve pentagonal and four hexagonal faces, E_{28}, creating a center with $\bar{4}3m$ symmetry. The cubic unit cell has a space group of $Fd\bar{3}m$. The type II compound can be formed nonstoichiometrically, without all of the "cages" filled. In general, the average

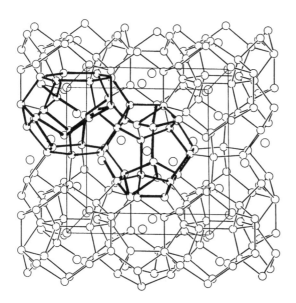

Fig. 6.12. The type I clathrate crystal structure. Only the group-IV elements are shown. Outlined are the two different polyhedra that form the unit cell. Reprinted from [6.53]. Copyright 2000, American Physical Society

interatomic distances are slightly larger than that of the analogous diamond-structured compounds. The average IV–IV–IV bond angles range from 105° to 126°, and average close to 109.5° that is characteristic of the tetrahedral angle in the diamond structure. The volume per fourth column atom of these two clathrate compounds, however, is larger than those of their analogous diamond-structured compounds (~ 15%). This is a good indication of the openness of these "open structures".

Nolas and co-workers [6.39, 6.49–6.51, 6.55, 6.56] extensively studied the structural and transport properties of type I germanium clathrates for potential thermoelectric applications. From the extensive structural analyses on these compounds, an estimate of the size of the polyhedra as a function of the entrapped "guest" can be obtained. Schujman et al. [6.50] compared the structural properties of type I Ge clathrates with an "ideal" Ge_{46} clathrate. This comparison illustrated that introducing the filler atoms into the structure expands it slightly with respect to the ideal empty structure; however, once the cages are filled, the size of the filler atom does not influence the size of the tetrakaidecahedral cages. The pentagonal dodecahedral cages expand slightly upon filling. This is quite interesting and a positive result for thermoelectric applications because it suggests that appropriate "guest" atoms incorporated into the polyhedra might further reduce λ_L toward λ_{min}.

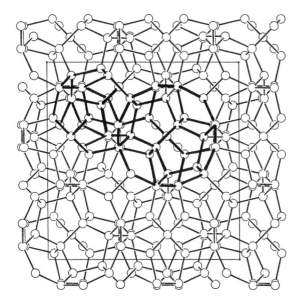

Fig. 6.13. The type II clathrate crystal structure. Outlined are the two different polyhedra that form the unit cell; only the group-IV elements are shown

There have been several published band-structure calculations on C [6.57, 6.58], Si [6.41. 6.43, 6.59–6.63], and Ge [6.61, 6.64–6.66] type I and II clathrates; however, none to date on Sn clathrates. In general these calculations showed that both clathrate types have indirect band gaps that are 0.7 eV wider than that of the diamond-structured phase whereas the volume in the clathrate phase is 17% larger.

Type I and II carbon clathrates result in similar volume expansion but have larger energies compared to diamond. Demkov et al. [6.62] also investigated the effect of alkali metals inside the polyhedra of Si clathrates. The electronic properties of the Si_{136} clathrate change, and perhaps may be altered in a controllable fashion by doping with alkali metals. The effect of doping on the band structure is stronger for less electropositive metals. In particular, doping with Na introduces a narrow band that comes from the lowest conduction band in pure Si_{136}. This band is half occupied at low Na concentration (x = 4 in Na_xSi_{136}) and is fully occupied at higher concentrations (x = 8). For K doping the effect is less pronounced, and the lowest conduction bands will have structures similar to that of pure Si_{136}.

Dong et al. [6.61] investigated the electronic structure and vibrational modes of pure Ge_{46} and Si_{136} and reported energies of 0.04 to 0.05 eV per atom higher than that of diamond structured Ge, whereas their volumes are 13 to 14 % higher. Both type I and II Ge clathrates are semiconductors with calculated gaps 1.21 and 0.75 eV wider than diamond-Ge, respectively; however the authors estimate that the true band gaps are near 2 eV due to the fact that their local density approximation (LDA) calculations normally underestimate these values. The phonon dispersion curves show that the acoustic modes are limited to less than 60 cm^{-1}. The Raman and infrared optic modes were also calculated. Zhao et al. [6.65] studied the structure and electronic properties of Ge_{46} and K_8Ge_{46} employing first principle calculations within the local-density approximation. These calculations indicated that Ge_{46} is a stable structure with only a slightly higher energy compared to diamond-structured Ge. The calculated band gap and expanded volume of Ge_{46} were in agreement with those obtained by Dong et al. [6.66]. K_8Ge_{46}, however, is metallic and has a moderate density of states. The valence band and density of states (DOS) are similar to those of pure Ge_{46}, whereas the conduction bands are modified by K. Almost complete charge transfer takes place from the K sites to the Ge framework.

6.3.4 Atomic Rattlers

As mentioned before, the specific framework and thermal parameters associated with the atoms in these compounds are an important aspect of this structure and have an effect on transport properties. One of the more unique and interesting structural features is the thermal motion of the "guest" atoms inside their atomic "cages". As illustrated in Fig. 6.10, the "guests" occupy two distinct sites. These sites are typically centered inside the polyhedra formed by the framework atoms that are comprised of three distinct sites. From room temperature structural refinements, the anisotropic atomic displacement parameters (ADPs) for Sr(2), obtained from single crystal and powder $Sr_8Ga_{16}Ge_{30}$, are enormous compared with those of the other atomic positions [6.44, 6.49, 6.51]. The difference in the ADPs between the Sr(2) atoms and the Sr(1) atoms is illustrated in the schematic shown in Fig. 6.14(a). The Sr(1) site has a more symmetric ADP and is slightly larger than that of the (Ga,Ge) framework atoms. The Sr(2) site exhibits an anisotropic ADP that is almost an order of magnitude larger than those of the other constituent atoms. Relatively large and anisotropic ADPs are typically observed for relatively small ions in the tetrakaidecahedra of this crystal structure. The enormous ADP for the Sr(2) site in $Sr_8Ga_{16}Ge_{30}$ implies the possibility of a static disorder in addition to the dynamic, or "rattling" motion. The electrostatic potential within the polyhedra is not the same everywhere, and different points may be energetically preferred. Moreover, anisotropic refinement of the displacement parameters indicates a much smaller displacement amplitude in the <100> directions than in the perpendicular directions [6.49, 6.51], as seen in (6.14a). This suggests that the Sr(2) position at 1/4,1/2,0 could also be described by splitting it into four positions within the {100} planes, as shown in Fig. 6.14(b). This would suggest that the atoms in the Sr(2) site can tunnel between the different energetically preferred positions. The possibility of a "freeze-out" of the "rattling" motion of Sr^{2+} in $Sr_8Ga_{16}Ge_{30}$ was originally indicated by low temperature λ_L data [6.55]. The implication is that static disorder is associated with a spatial distribution of the Sr(2) positions inside the polyhedra. It is plausible that the large ADPs measured contain both a static as well as a dynamic component. This was best illustrated by Chakoumakos et al. [6.51] from refinements of neutron diffraction data on single crystal and powder $Sr_8Ga_{16}Ge_{30}$, as shown in Fig. 6.15.

In Fig. 6.15, the temperature dependence of the isotropic ADP, $U(T)$, for the atomic positions in $Sr_8Ga_{16}Ge_{30}$ in the split-site (static + dynamic disorder) model as well as in the single-site (dynamic disorder) model is shown [6.51]. At room temperature, $U(T)$ is large for Sr(2) in the single-site model, whereas the low-temperature intercept implies a large static positional disorder at absolute zero. With a model in which the Sr(2) fractionally occupies a multiply split site, the $U(T)$ values are dramatically reduced but still remain the largest of all of the atoms in the crystal. The split-site model implies less thermal disorder. Unconstrained refinement of the Sr(2) site occupancy indicates no deficiency in occupancy of the Sr(2) position for either model. The ADP for the Sr(1) site is also large but more on par with the (Ga,Ge) framework sites (indicated by M in Figure 6.15) due to its smaller sized cavity. The positions of the framework (Ga,Ge) atoms show a minimal temperature dependence, which indicates a relatively stiff framework. This is consistent with a model whereby the Sr atoms "rattle" about in the relatively large tetrakaidecahedron "cage".

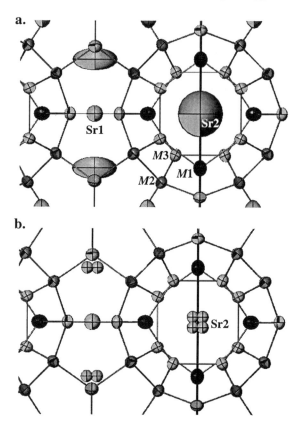

Fig. 6.14. Crystal structure projection on a (100) plane of $Sr_8Ga_{16}Ge_{30}$ illustrating the large anisotropic atomic displacement parameters for the single-site model (**a**), and the split-site model (**b**) where the combined static and dynamic disorder of the Sr(2) crystallographic site with isotropic atomic displacement parameters is indicated. Reprinted from [6.51] with permission from Elsevier Science

Nolas and co-workers [6.49, 6.51, 6.56, 6.67] employed the split-site or off-center, multiple-well model, obtained from the structural refinements, described in the previous section, in evaluating the λ_L data. The model incorporated dynamic as well as static disorder associated with the encapsulated atoms, as required. The ADP data were also employed in determining the characteristic localized vibrational frequencies for weakly bound atoms that "rattle" within their atomic "cages". This approach, which assumes that the "rattling" atoms act as harmonic oscillators, was successful in systems that contain a dynamic disorder in an otherwise rigid framework structure as shown by Sales et al. [6.68].

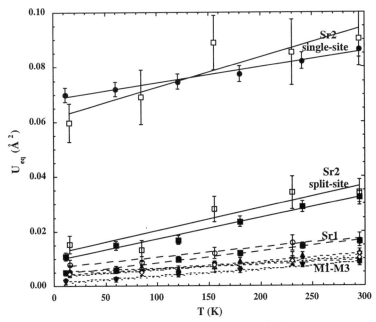

Fig. 6.15. Temperature dependence of the isotropic displacement parameters measured on single-crystal (open symbols) and powder (closed symbols) $Sr_8Ga_{16}Ge_{30}$. The symbols M represent the three Ge sites. Reprinted from [6.51] with permission from Elsevier Science

As revealed by this semiquantitative analysis, at low temperatures, the static disorder associated with the Sr positions in the tetrakaidecahedra, described by the split-site model for the Sr(2) site, may be the origin of the tunneling system (TS) associated with a "freezing-out" of the "rattling" motion of Sr(2). This would thereby result in the T^2 temperature dependence, shown in Fig. 6.11. The fit to the λ_L data resulted in TS parameters that are comparable to those for many amorphous solids [6.49, 6.55, 6.67]. Small energy barriers presumably characterize the separation between these Sr(2) positions. The resonance dip in the data in the range $4 < T < 35$ K is associated with the dynamic, or "rattling", of the Sr atoms inside their atomic cages (Fig. 6.11).

The correlation between the crystal structure and the thermal conductivity of semiconducting clathrates can be expanded to include other clathrate compounds, as well as defect structured compounds, as described by Nolas et al. [6.46, 6.53, 6.54, 6.69]. Although the enhanced thermal motion of the "guest" atoms inside their atomic "cages" in the Si and Sn clathrates appears to diminish λ_L, the effect is not as great as that caused by Sr(2) (Eu(2)) motion in $Sr_8Ga_{16}Ge_{30}$ ($Eu_8Ga_{16}Ge_{30}$). There the Sr(2) and Eu(2) thermal ellipsoids are nearly an order of magnitude larger than those of the (Ga,Ge) framework atoms. It may be that static, as well as dynamic disorder, is required to achieve glass-like thermal transport in these compounds.

6.3.5 Thermal Conductivity

The measurement and analysis of the thermal conductivity of clathrates has to date mainly been undertaken by Nolas and co-workers [6.39, 6.46, 6.49, 6.53–6.55, 6.67, 6.69]. Figure 6.16 shows λ_L for the two crystalline Ge clathrates in the range $5 < T < 100$ K [6.49]. Also shown is λ of vitreous silica (a-SiO$_2$). The solid line fits to the data were obtained by employing an analysis similar to that described in the previous section. The "resonance dips" in these single-crystal Ge clathrate specimens are much more pronounced than the polycrystalline specimens, shown in Fig. 6.16, due to the lack of grain-boundary scattering from micron-sized polycrystalline grains. Eu$_8$Ga$_{16}$Ge$_{30}$ has a lower λ_L at all temperatures and a more pronounced "dip" due to the fact that the Eu^{2+} ion will have a larger effect on λ_L than Sr^{2+} because Eu^{2+} is more massive. The Eu(2) ADP is also slightly larger than that of Sr(2) in Sr$_8$Ga$_{16}$Ge$_{30}$ [6.49]. A similar resonance scattering of phonons is the principal mechanism that determines low λ_L values in lanthanide-filled skutterudites (Sect. 6.2.2).

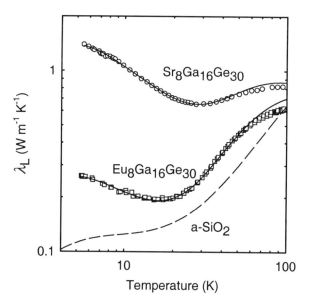

Fig. 6.16. Lattice thermal conductivity vs. temperature of single-crystal Sr$_8$Ga$_{16}$Ge$_{30}$ and Eu$_8$Ga$_{16}$Ge$_{30}$ along with that of amorphous SiO$_2$. Reprinted from [6.49]. Copyright 2000, American Physical Society

In Fig. 6.17, λ_L from room temperature to 6 K is shown for five representative clathrate specimens [6.55]. From this figure a trend in the thermal transport emerges that is intimately related to the crystal structure of these compounds. The Ge clathrates exhibit a λ_L that is typical of amorphous solids. The values are well below that of a-SiO$_2$ and close to a-Ge near room temperature. In addition, the temperature dependence of the thermal conductivity of Ge clathrates is much like that of a-SiO$_2$. Europium (Eu^{2+}) is a smaller ion and has a larger ADP than Sr^{2+} in the (Ga,Ge) framework [6.49]. In addition, Eu^{2+} is almost twice as massive as Sr^{2+} and therefore, will have a larger effect on λ_L. In the case of Sr$_4$Eu$_4$Ga$_{16}$Ge$_{30}$, there are two different atoms in the voids of the crystal structure which introduce six different resonant scattering frequencies (three for each ion). This compound exhibits the lowest λ_L values in the temperature range shown and tracks the temperature dependence of a-SiO$_2$ quite closely. The λ_L measurements of Sr$_4$Eu$_4$Ga$_{16}$Ge$_{30}$ clearly demonstrate that λ_L is further reduced when the disorder of Sr or Eu is solely employed. We note that in the temperature range below 30 K,

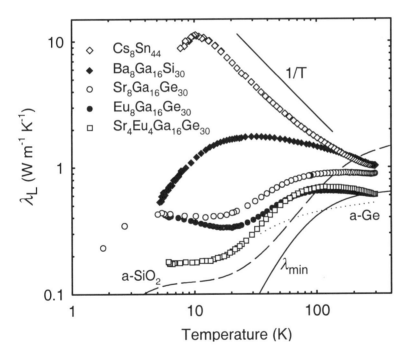

Fig. 6.17. Lattice thermal conductivity vs. temperature for five representative polycrystalline type I clathrates. The dashed and dotted curves are for amorphous SiO$_2$ and amorphous Ge, respectively, and the solid curve is λ_{min} calculated for Ge. Reprinted from [6.55]. Copyright 2000, American Physical Society

there is also grain-boundary scattering of the phonons, and single-crystal samples have somewhat higher λ_L values in this temperature range [6.49, 6.68]. The $Ba_8Ga_{16}Si_{30}$ sample has a relatively low λ_L; however, the temperature dependence is not similar to that of Ge clathrates. Although Ba is much more massive than the elements that make up the host matrix (i.e., Ga and Si), a prerequisite for glass-like λ_L [6.1], the temperature dependence of λ_L does contain a somewhat crystalline character because Ba^{2+} is similar in size to the Si_{20} and Si_{24} cages, whereas Sr^{2+} and Eu^{2+} are smaller than the Ge cages. From X-ray diffraction data, it does not appear that Ga increases the average size of the cage very much. Cs_8Sn_{44} exhibits a T^{-1} temperature dependence typical of simple crystalline insulators dominated by phonon–phonon scattering. The low λ_L values may be due to the complex crystal structure of fifty-six atoms per unit cell in this compound. The additional bonding induced between the Cs and Sn atoms neighboring the vacancies in Cs_8Sn_{44} apparently constrains the Cs(2) displacements. This structural difference between Cs_8Sn_{44} and that of the other clathrates appears to be the source of their differing thermal conductivities.

6.3.6 Electronic Properties

In the early work of Cros and co-workers [6.70] the electronic properties of several of their Si clathrates were investigated. Figure 6.18 shows the Seebeck coefficient for four of their n-type specimens, three type-II Si clathrates along with that for K_7Si_{46}, prepared by cold pressing and then sintering at 400°C [6.70]. The Seebeck coefficient increases with decreasing Na content. This is in agreement with their magnetic susceptibility data and indicates a metal-to-insulator transition at x ~ 11 in Na_xSi_{136}. Their electronic conductivity data, however, were compromised by the quality of the specimens prepared for electronic measurements [6.71]. A thorough investigation of the transport properties of the type II clathrate system is yet to be undertaken.

Figure 6.19 shows α and ρ as functions of temperature from 300 to 5 K for four specimens with nominal compositions $Sr_8Ga_{16+x}Ge_{30-x}$, where x varied slightly in the three specimens [6.39]. The ability to vary the doping level of these semiconductor compounds by varying the chemical composition is one reason why they are of interest for thermoelectric applications. Ga is randomly substituted for Ge in the structure and is used to produce charge compensation for the divalent alkaline-earth ion Sr^{2+}. The doping level of this series of specimens was varied by changing the Ga to Ge ratio while maintaining a fixed Sr concentration. This is similar to doping diamond-structure Ge with As, for example. p-type conduction was not obtained in the clathrate in this manner; however, p-type clathrates can be achieved by substituting Zn for Ge [6.72]. The magnitude of α decreases with

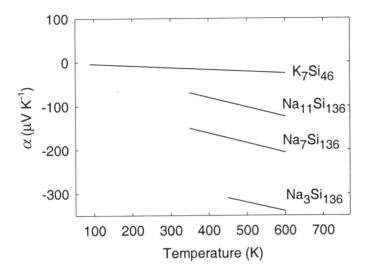

Fig. 6.18. Seebeck coefficient of three type II alkali-metal Si clathrates plus K_7Si_{46}. Reprinted from [6.70] with permission

increasing carrier concentration, as shown in Fig. 6.19(a). The magnitude of α also decreases with decreasing temperature, as expected in heavily doped semiconductors with negligible phonon drag. The highest room temperature mobility of these samples was of the order of 100 cm^2 V^{-1} s^{-1}. The intrinsic carrier mobility in the clathrate structure is unknown at present. The presence of ionized electron donors such as Na^{+1} or Sr^{+2} decreases the mobility below its intrinsic value, but by exactly how much is not yet fully understood.

Nolas and co-workers measured the electronic transport properties of several type I Si, Ge, and Sn clathrate specimens. Compounds exhibited either semiconducting or metallic electronic properties, depending on doping level and, in certain compounds, on stoichiometry. In general, α increases with increasing temperature in these compounds, typical of semiconductor behavior, and the magnitude depends on the carrier concentration. Very high α values and very high electrical resistivities were measured in Sn clathrates. N-type conduction is most often obtained in clathrate compounds; however, p-type compounds have also been synthesized [6.72]. In certain compositions such as $Sr_8Zn_8Ge_{38}$, n- and p-type compounds can be synthesized [6.72].

The fact that clathrate compounds can be synthesized so that they possess glass-like lattice thermal conductivity, the ability to vary the electronic properties by changing the doping level, and relatively good electronic properties indicate that this system is a PGEC, and therefore it is of interest for further research for thermoelectric applications. This is the only known system of compounds that has relatively good semiconductor properties in addition to exhibiting glass-like λ_L properties. In these compounds, the goal is to replace the traditional alloy phonon scattering, which predominantly scatters the highest frequency phonons, with a much lower frequency resonance and possibly disorder-type scattering. The highest frequency phonons in any crystal lattice have very low or zero group velocities and contribute little to the total thermal conductivity. The low-frequency phonons have the highest group velocity and contribute most to λ_L. This is why the "caged impurities" produce such pronounced decreases in λ_L in clathrate and cryptoclathrate systems.

Thus far, the highest room temperature dimensionless figure of merit zT for clathrate compounds ranges from 0.25 to 0.34. The zT values increase with increasing temperature, and $zT > 1$ above 400°C. These values were obtained on polycrystalline type-I Ge clathrates. These results are very interesting for thermoelectric applications and represent a high zT value for new and unoptimized materials. Presumably, the zT values could be increased by optimizing the doping level, changing the rattle frequencies, or by employing single-crystal samples instead of polycrystalline ones. The promising properties of these interesting substances suggest a new category of thermoelectric materials for continued research and investigation.

7. Complex Chalcogenide Structures

7.1 Introduction

Traditionally, the use of high temperatures to directly combine elements or simpler compounds into more complex ones has been quite successful in providing new materials. However, this approach does give rise to important synthetic limitations. For example, the reactions almost always proceed to the most thermodynamically stable products; the high energies involved often leave little room for kinetic control. These thermodynamically stable products are typically the simplest binary or ternary compounds, and because of their high lattice stability, they become synthetic obstacles. Second, the high reaction temperatures also dictate that only the simplest chemical building blocks can be used, that is, elements on the atomic level. Synthetic attempts using molecules of known structure are doomed because the high temperatures reduce the system to a thermodynamic minimum, thereby not alloying for the desired bond formation. Hence, multinary compounds can be more difficult to form, and the preference lies with more stable binary and ternary compounds. Being almost totally at the mercy of thermodynamics, the solid state chemist has traditionally relied on experience and intuition, rather than a set of predictable rules.

One approach intended to overcome this difficulty was outlined in Chapter 6. Another approach has been to perform reactions using molten salts as solvents. Such media have been employed for more than 100 years for high-temperature, single-crystal growth [7.1]. Although many salts melt at high temperatures, eutectic combinations of binary salts and salts of polyatomic species often have melting points well below the temperatures of classical solid-state synthesis, making possible their use in exploring new chemistry at intermediate temperatures. In many cases, such salts act as solvents and also as reactants, providing species and building blocks that can be incorporated into the final product. In the search for new thermoelectric materials, Kanatzidis and co-workers [7.2–7.7] demonstrated that this molten salt method is most suitable for exploratory synthesis involving heavy elements such as Ba, Sr, Bi, Pb, Sn, Se, Te, and alkali-metal atoms. Because chalcogenide compounds are presently used in thermoelectric devices, Kanatzidis and co-workers began a search for more complex semiconducting compounds of this type.

Structural and compositional complexity can result in corresponding complexities in electronic structure that may produce beneficially large asymmetry in the density of states (DOS) to obtain large values of the Seebeck coefficient α. The phonon contribution to thermal conductivity λ can also be reduced by such structural complexity, by choosing heavy elements, and by choosing combinations of elements that make moderate to weak chemical bonds.

The concept that certain materials can conduct electricity like a crystalline solid but heat like a glass (the PGEC approach discussed in Chapter 6) is also employed here. In such materials, a weakly bound atom or molecule can create localized dynamic disorder, or "rattle", thereby resulting in a low λ value for the solid without severely affecting electronic conduction and potentially leading to improved thermoelectric materials. The class of chalcogenide materials described here may also satisfy this description because, as will become apparent later, they are made of two- or three-dimensional bismuth chalcogenide, group-IV chalcogenide, or chalcogenide frameworks stabilized by weakly bonded alkali-metal atoms which reside in cavities or tunnels in the framework. These electropositive atoms almost always possess the highest atomic displacement parameters (ADPs) in the structure, which is evidence that a certain degree of disorder may be present. This feature is very important in substantially suppressing λ in these materials. These materials also incorporate other interesting features such as large unit cells and complex compositions that are beneficial for potentially improving thermoelectric properties.

7.2 New Materials with Potential for Thermoelectric Applications

The fact that alloys of Bi_2Te_3 and Bi–Sb are the best thermoelectric materials known to date below 100°C suggests that they possess features necessary for high figures of merit. Therefore, Kanatzidis and co-workers [7.2–7.7] reasoned that similar properties may manifest in similar compounds. Their research direction is to explore other multinary chalcogenides of bismuth and antimony in the hope that these elements will impart some (or all) of the key properties needed for superior thermoelectric materials. Based on these considerations, Kanatzidis and co-workers decided to perform exploratory chemical synthesis involving Bi as one of the elements. As outlined in this section, promising new materials have been found. The other elements employed in the synthesis are chalcogens such as S, Se, and Te, as well as alkali metals.

The presence of alkali metals in the structures of ternary and quaternary bismuth chalcogenides induces the stabilization of covalently bonded Bi chalcogen frameworks with cages or tunnels that accommodate the charge-balancing alkali atoms. The interactions of alkali metals and the Bi chalcogen frameworks are considered mainly electrostatic. Alkaline-earth metals such as Sr and Ba (and by extension the lanthanide Eu) with a 2+ charge are also desirable as substitutes for alkali metals. They tend to have properties similar to alkali metals, but their greater charge tends to stabilize different structures.

The considerations outlined before led these researchers to pseudoternary chalcogenide compounds constructed from simple building blocks, which can be regarded as fragments excised from the basic PbS lattice. These fragments formally derive from "cutting" the NaCl structure along various directions that generate blocks of various sizes and shapes. The theme in all structures seems to be that these building blocks vary in size and width, and they are usually infinite in at least one dimension. In addition to size and width, the particular "cut" (i.e., crystallographic orientation) is also important and varies from compound to compound.

7.2.1 Approach to Synthesis

To discover chalcogenide materials, the use of molten alkali-metal polychalcogenides of type A_2Q_x (A = alkali metal, Q = S, Se, Te) as solvents is very appropriate, as previously demonstrated [7.8, 7.9]. A_2Q_x salts are especially well-suited as solvents for intermediate temperature reactions because the melting points range between 200 and 600°C. Most alkali polytelluride salts melt between 300 and 500°C. Low melting A_2Q_x fluxes remain nonvolatile over a wide temperature range, and so, once above the melting point, reaction temperatures can be varied considerably without concern for solvent loss. Polychalcogenide fluxes are highly reactive toward metals because they are strong oxidants. Reactions between metals and molten A_2Q_x are performed *in situ*. The powdered reagents (polychalcogenide and metal or metal chalcogenide) are mixed under inert atmosphere and loaded into pyrex or silica reaction vessels. Once evacuated, the tubes are sealed under vacuum and subjected to the desired heating program in a computer-controlled furnace. In a given system, large compositional ranges are explored simultaneously (many synthesis runs in a given time) so that all possible phases can be found. In this sense, this is a "poor man's" combinatorial method. Because this is a flux technique, only the part of the phase space that is rich in flux can be explored. Metal-rich compositions will have to be investigated in other ways.

To synthesize new compounds, one or more metals are added directly to the molten A_2Q/Q reaction mixture and heated in a sealed pyrex or silica container. Crystalline products either precipitate from the melt or form on slow cooling of the melt, depending on the specific stoichiometric and processing conditions. Presumably, the nucleated species are in equilibrium with the soluble intermediates, especially if the flux is present in excess, and hence a solvation/reprecipitation effect (often referred to as the mineralizer effect) occurs. This aids in growing single crystals because the flux can redissolve small or poorly formed crystallites and then reprecipitate the species onto larger, well-formed crystals. The advantage of the flux method is that one allows the system to end up "where it wants" in the kinetic or thermodynamic sense without attempting to force a certain stoichiometry or structure upon it. Provided that the temperature and time are appropriate, the reaction systems have all of the ingredients and freedom to form new phases. The benefit of this becomes apparent from the unusual compositions often found in the new materials that most certainly could not have been predicted a priori.

212   7. Complex Chalcogenide Structures

The molten flux approach is suitable for quick and broad explorations of phase space but is most practical when carried out in small (< 2 g) quantities. In this sense, it is a discovery approach, not a production method. Of course, once a new material has been discovered by such techniques, the next step is to attempt to large-scale preparation. Therefore, once the composition of a new material has been established, other synthetic methods amenable to scale-up must then be employed.

7.2.2 The selenides β-$K_2Bi_8Se_{13}$ and $K_{2.5}Bi_{8.5}Se_{14}$

The compounds β-$K_2Bi_8Se_{13}$ (isostructural to $K_2Bi_8S_{13}$) and related $K_{2.5}Bi_{8.5}Se_{14}$ are very stable. As found in the isostructural sulfide, the high coordination sites in the lattice (i.e., those with coordination number 7 or higher) are occupied by both K^+ and Bi^{3+} ions. The octahedral Bi sites have a smaller cavity size and do not accept K^+ ions. Both are three-dimensional, but β-$K_2Bi_8Se_{13}$ has a more open structure than $K_2Bi_8S_{13}$. Overall, the structure of $K_2Bi_8S_{13}$ is slightly more dense than that of the β-form, because 25 % of the Bi atoms in the latter are found in a trigonal pyramid, and in the former all Bi atoms are in an octahedral or higher coordination geometry (Fig. 7.1).

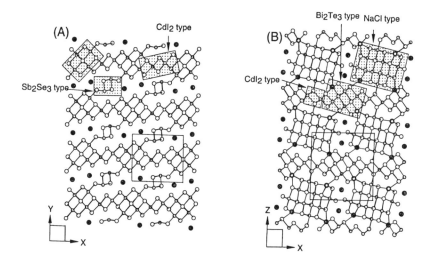

Fig. 7.1. (A) Projection of the structure of β-$K_2Bi_8Se_{13}$ viewed down the b axis; NaCl, Bi_2Te_3, and CdI_2-type fragments in this framework are highlighted. (B) Projection of the structure of α-$K_2Bi_8Se_{13}$ viewed down the b axis; Sb_2Se_3, Bi_2Te_3, and CsI_2-type building blocks are highlighted. Reprinted from [7.4]. Copyright 1997, American Chemical Society

7.2 New Materials with Potential for Thermoelectric Applications 213

$K_{2.5}Bi_{8.5}Se_{14}$ possesses a complex three-dimensional anionic framework related to that of β-$K_2Bi_8Se_{13}$, as shown in Fig. 7.2. Compositionally, $K_{2.5}Bi_{8.5}Se_{14}$ derives from β-$K_2Bi_8Se_{13}$ by an addition of 0.5 $KBiSe_2$. The main difference between the two structures is that only NaCl- and Bi_2Te_3-type blocks exist in $K_{2.5}Bi_{8.5}Se_{14}$. $K_{2.5}Bi_{8.5}Se_{14}$ forms by addition of half "$BiSe_2$" atoms to a CdI_2-type fragment in β-$K_2Bi_8Se_{13}$. This small structural modification preserves the same connectivity of the NaCl-type fragments as well as the size and shape of this K site as in β-$K_2Bi_8Se_{13}$. Although the width of the NaCl block in the structure of $K_{2.5}Bi_{8.5}Se_{14}$ is also that of three Bi polyhedra, its Bi_2Te_3 block is five Bi polyhedra wide. This is an important difference between this selenide and β-$K_2Bi_8Se_{13}$ (see Figs. 7.1 and 7.2).

The electrical properties of β-$K_2Bi_8Se_{13}$ and $K_{2.5}Bi_{8.5}Se_{14}$ were measured from undoped single-crystal samples and polycrystalline ingots. Typical room temperature values of approximately -200 $\mu V\ K^{-1}$ and 250 $\Omega^{-1}\ cm^{-1}$ were obtained for α and electronic conduction σ, respectively, and a weak negative temperature dependence of σ was consistent with a semimetal or a narrow band-gap semiconducting material. In general, $K_{2.5}Bi_{8.5}Se_{14}$ is obtained from the synthesis in a more highly doped state than β-$K_2Bi_8Se_{13}$ because its α value tends to be smaller and its electronic conductivity greater. β-$K_2Bi_8Se_{13}$ and $K_{2.5}Bi_{8.5}Se_{14}$ are narrow band-gap semiconductors whose room temperature band gaps are 0.59 and 0.56 eV, respectively. Greater band gaps were found in Sb analogs; 0.78 eV for

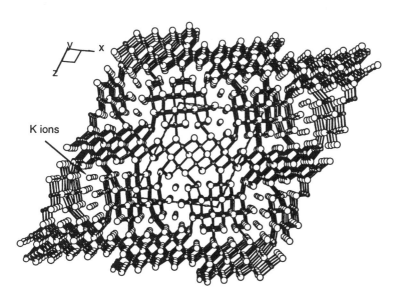

Fig. 7.2. Structure of $K_{2.5}Bi_{8.5}Se_{14}$ projected down the *b* axis

$K_2Sb_8Se_{13}$ and 0.82 eV for $K_{2.5}Sb_{8.5}Se_{14}$. Tunability in the band gaps and consequently in electrical properties could be achieved by preparing solid solutions of the type $K_2Bi_{8-x}Sb_xSe_{13}$ and $K_{2.5}Bi_{8.5-x}Sb_xSe_{14}$. Due to the narrower band gap of $K_{2.5}Bi_{8.5}Se_{14}$ better properties can be obtained, compared to β-$K_2Bi_8Se_{13}$, provided that appropriate doping is achieved.

The thermal conductivities of both β-$K_2Bi_8Se_{13}$ and $K_{2.5}Bi_{8.5}Se_{14}$ are very low: 1.28 and 1.24 W m^{-1} K^{-1}, respectively, at room temperature. These values are comparable to those of Bi_2Te_3 alloys. The low λ values are attributed to a number of factors, including the low crystal symmetry (monoclinic), large unit cells, and the presence of "rattling" alkali atoms in tunnels, which interact only electrostatically with Se atoms on the tunnel walls.

7.2.3 Doped β-$K_2Bi_8Se_{13}$ and $K_{2.5}Bi_{8.5}Se_{14}$

β-$K_2Bi_8Se_{13}$ doped with varying concentrations of $SbBr_3$ and Sn was synthesized by Kanatzidis and co-workers. The σ values for the Sn-doped samples generally decreased as the doping level increased, as shown in Fig. 7.3. The room temperature σ values increased when the material was doped with 0.5% Sn, but conductivities steadily decreased as the doping concentration increased from 0.5 to 3.0% Sn. The temperature dependence also showed a transition from weakly metallic behavior to semiconducting behavior. Except for the 2.0% Sn sample, which showed metallic behavior, the σ data suggest that the activation energy increases with doping level. Temperature-dependent α data are shown in Fig. 7.4. Based on the approximate relationship $E_g \sim 2\alpha_{max}T_{max}$ [7.10] and the optically measured $E_g \sim 0.54$ eV, α is expected to maximize at ~500 μV K^{-1} in the region between 500–700 K. This temperature range is also the range of maximum zT for this material. The α values showed a steady increase with doping concentration, and the room temperature value steadily increased from –220 μV K^{-1} for the undoped β-$K_2Bi_8Se_{13}$ to –333 μV K^{-1} for the sample doped with 3.0% Sn.

The power factor $\alpha^2\sigma$ for varying dopant concentrations is plotted as a function of temperature in Fig. 7.5. The power factor data are a good indication of the optimal doping level for each dopant, as well as the temperature where $\alpha^2\sigma$ is maximized. The most promising sample was 0.5% Sn-doped β-$K_2Bi_8Se_{13}$, whose maximum $\alpha^2\sigma$ was 38.5 μW cm^{-1} K^{-2} at 295 K. This value is approximately a three-fold increase from that of undoped β-$K_2Bi_8Se_{13}$ at room temperature. A further increase in doping levels resulted in a steady reduction of $\alpha^2\sigma$.

The $\alpha^2\sigma$ values for the $SbBr_3$-doped specimens indicated a sharp drop with doping level to 1.0%, then an increase to the highest values for the 2.2% doped sample. The $\alpha^2\sigma$ values then decreased as doping increased to 3.0%. Although the enhancement of $\alpha^2\sigma$ was not as dramatic as that for the 0.5% Sn-doped sample, the 2.2% $SbBr_3$-doped sample still exhibited a substantial increase, compared to undoped β-$K_2Bi_8Se_{13}$.

7.2 New Materials with Potential for Thermoelectric Applications   215

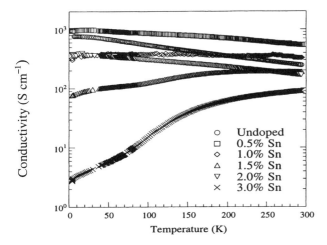

Fig. 7.3. Electrical conductivity as a function of temperature for Sn-doped β-$K_2Bi_8Se_{13}$

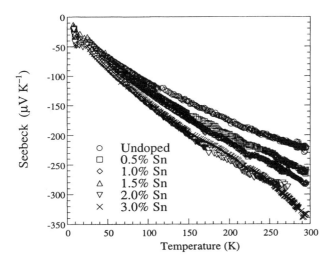

Fig. 7.4. Seebeck coefficient versus temperature for Sn-doped β-$K_2Bi_8Se_{13}$

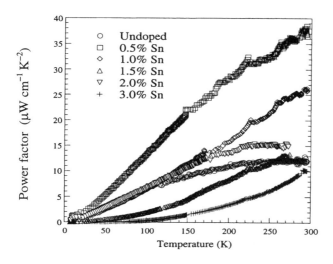

Fig. 7.5. Power factor versus temperature for Sn-doped β-$K_2Bi_8Se_{13}$

It is apparent from these studies that $\alpha^2\sigma$ exceeds that of undoped β-$K_2Bi_8Se_{13}$ by approximately a factor of 3 with a relatively small increase in λ. Greater enhancements are required, however, for these interesting materials to prove useful in thermoelectric cooling applications. A reasonable approach to improve z of these ternary compounds could be to form solid solutions of $K_xBi_y(Se,S)_z$ and $K_x(Bi,Sb)_ySe_z$. This type of alloying may decrease λ further and perhaps increase α by changing the band gap. An outstanding challenge is to achieve p-type doping in these materials. Although such efforts are beginning, so far, only n-type transport has been observed.

7.2.4 CsBi$_4$Te$_6$

In sharp contrast to the selenide compounds discussed in the previous section, the isostructural Te analogs are not stable. Rather, the A/Bi/Te and A/Pb/Bi/Te systems, where A represents an alkali metal, give rise to different structural types. One of the more interesting of them is CsBi$_4$Te$_6$. From a chemical point of view, this compound represents a reduction of a Bi$_2$Te$_3$ unit by a half equivalent of electrons. This reduction results in a complete restructuring of the Bi$_2$Te$_3$ framework, so that a new structure forms. This compound is of interest for low-temperature thermoelectric applications [7.7]. The material is air- and water-stable and melts without decomposition at 545°C. The crystals grow with a long needle-like morphology. The direction of rapid growth along the needle axis is also the direction of maximum thermoelectric performance.

7.2 New Materials with Potential for Thermoelectric Applications

CsBi$_4$Te$_6$ has a layered anisotropic structure. As shown in Fig. 7.6, it is composed of anionic Bi$_4$Te$_6$ slabs alternating with layers of Cs$^+$ ions. The added electrons localize on the Bi atoms to form Bi–Bi bonds that are 3.238 Å long. The presence of these bonds is unusual in bismuth chalcogenide chemistry, and it is not clear whether they play a role in the thermoelectric properties of the material [7.11]. The strongly anisotropic Bi$_4$Te$_6$ layers consist of one-dimensional (1-D) Bi$_4$Te$_6$ ribbons running parallel to the b axis. The width and height of these rods are 23 Å by 12 Å; see Fig. 7.7. The ribbons arrange side by side and are connected via the Bi–Bi bonds mentioned before. This structural feature is responsible for the strongly 1-D needle-like appearance of the CsBi$_4$Te$_6$ crystals. The Bi atoms are octahedrally surrounded by either six Te atoms or by five Te atoms and one other Bi atom. The degree of distortion around the Bi atoms is relatively small. The longest and shortest Bi–Te bonds are 3.403(1) Å and 2.974(1) Å, respectively, and the average distance is 3.18 Å. The Cs$^+$ ions lie between the Bi$_4$Te$_6$ layers, and their ADPs are 1.6 times greater than those of the Bi and Te atoms, which suggests that they undergo considerable "rattling", or dynamic, motion. This dynamic motion in the lattice can be responsible for strong scattering of heat-carrying phonons and leads to low λ values. The immediate environment of Cs is a square prismatic arrangement of Te atoms.

As obtained directly from the synthesis (with no deliberate attempt at doping), crystals of CsBi$_4$Te$_6$ have metallic σ and room temperature α values of approximately 100 μV K^{-1}. The λ measurements on a large number of pressed pellets (> 97% theoretical density) show relatively low λ for this material; values at room temperature were approximately 1.5 W m^{-1} K^{-1}. This is comparable to that of Bi$_2$Te$_3$ alloys.

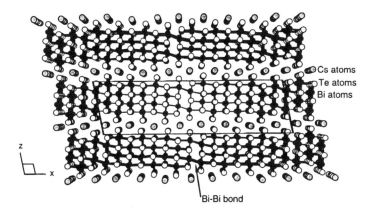

Fig. 7.6. The structure of CsBi$_4$Te$_6$ looking down the b axis. The Bi–Bi bond is indicated by an arrow. Reprinted from [7.7] with permission

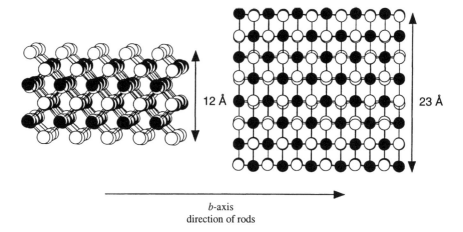

Fig. 7.7. Individual Bi_4Te_6 ribbons in the $CsBi_4Te_6$ structure viewed in two different orientations. The ribbons are joined side by side via Bi–Bi bonds. Reprinted from [7.7] with permission

Doping studies were undertaken with various chemical doping agents such as SbI_3, BiI_3, and In_2Te_3 in amounts varying from 0.02 to 4 mol% [7.7]. Doping with SbI_3 and BiI_3 produces p-doped samples, whereas In_2Te_3 produces n-type samples. Depending on the type and degree of doping, room temperature α values between +175 $\mu V\ K^{-1}$ and –100 $\mu V\ K^{-1}$ were observed. $CsBi_4Te_6$ is amenable to considerable doping manipulation, much like Bi_2Te_3, and thus improved zT values from doped $CsBi_4Te_6$ are of interest.

The best $\alpha^2\sigma$ values measured on single crystals were obtained with SbI_3. Figure 7.8(A) shows the evolution of the power factor as a function of SbI_3. From these data, the optimal concentration lies at 0.05% SbI_3. The temperature dependence of σ and α measured on the best sample are shown in Fig. 7.8(B). The α maximum was found at ~ 250 K. The combination of these measurements in obtaining zT for the 0.05% SbI_3-doped $CsBi_4Te_6$ with that of optimized $Bi_{2-x}Sb_xTe_3$ as a function of temperature is illustrated in Fig. 7.8(C). The reported zT values for $CsBi_4Te_6$ reach a maximum of 0.82 at 225 K and 0.65 at room temperature [7.7]. In contrast, zT of an optimized $Bi_{2-x}Sb_xTe_3$ p-type alloy peaks at ~ 0.95 at room temperature and at ~ 0.7 at 225 K. Because the $\alpha^2\sigma$ values of doped $CsBi_4Te_6$ reach a maximum well below that of $Bi_{2-x}Sb_xTe_3$, this new material is of interest for low-temperature applications.

Hall effect measurements for SbI_3-doped $CsBi_4Te_6$ samples show that carrier concentrations are on the order of 3×10^{18} to $10^{19}\ cm^{-3}$ for samples doped at 0.1% and 0.2% SbI_3, respectively. The hole mobilities in doped $CsBi_4Te_6$ samples range between 700 and 1000 $cm^2\ V^{-1}\ s^{-1}$ at room temperature. These are significantly greater than those typically found in p-type bismuth telluride alloys. At low temperatures, the mobility rises to > 5000 $cm^2 V^{-1} s^{-1}$. The very high hole mobilities

7.2 New Materials with Potential for Thermoelectric Applications 219

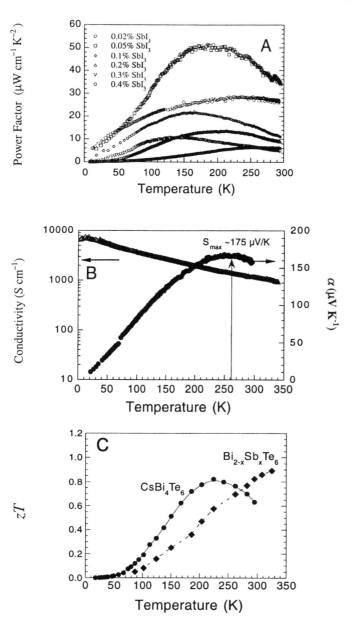

Fig. 7.8. Power factor (**A**) and electrical conductivity plus Seebeck coefficient (**B**) of SbI$_3$-doped CsBi$_4$Te$_6$ single crystal. (**C**) Figure of merit versus temperature for 0.05% SbI$_3$-doped CsBi$_4$Te$_6$. Also shown is Bi$_{2-x}$Sb$_x$Te$_3$. Reprinted from [6.32] with permission. Copyright 1996 American Association for the Advancement of Science

could be due to the 1-D character of $CsBi_4Te_6$ and the lack of atomic disorder in its crystal lattice.

A thermoelectric cooling device needs both p-type and n-type branches for operation. Thus, an issue to be addressed in future studies is whether good zT values can be found for n-type samples. Thus far Kanatzidis and co-workers demonstrated that In_2Te_3 doping leads to n-type charge transport. Though optimum levels have not yet been reached, a maximum α of $-100\ \mu V\ K^{-1}$ at ~160 K has been observed. If n-type zT remains low for $CsBi_4Te_6$-based materials, then there is likely to be renewed interest in developing durable $Bi_{1-x}Sb_x$ alloys to use with p-type $CsBi_4Te_6$ in the coldest stages of multistage modules.

Research into further improvements in this material is currently underway with continued exploration of doping agents and solid solutions such as $CsBi_{4-x}Sb_xTe_6$, $CsBi_4Te_{6-x}Se_x$ and $Cs_{1-x}Rb_xBi_4Te_6$. The latter could result in substantially lower thermal conductivity, and band-structure calculations for $CsBi_4Te_6$ should contribute to better understanding of this materials electronic properties [7.11].

7.3 Pentatelluride Compounds

7.3.1 Introduction

One of the more important issues related to developing low-temperature thermoelectric materials is identifying systems or mechanisms that might give high α values at low temperatures. Common systems or mechanisms include phonon drag, heavy fermion materials, Kondo systems, materials that exhibit phase transitions, as well as quasi-one-dimensional materials. Quasi-one-dimensional systems are known to be susceptible to van de Hove singularities (or cusps) in their density of states $g(E)$, electronic phase transitions, and exotic transport phenomena, which can add structure in $g(E)$ near the Fermi energy E_F. At temperatures far from a phase transition, ρ and α are related to the electron density of states near the Fermi energy $g(E_F)$. Conductivity is proportional to $g(E_F)$, and α is proportional to $(1/g)dg(E)/dE$ at $E=E_F$. Hence, as n (or g) is increased, σ typically increases, and α decreases. Doping can produce very substantial effects in these types of material and can drastically change their electronic transport. In this section, results on a family of low-dimensional semiconductors called pentatellurides ($HfTe_5$ and $ZrTe_5$) are discussed. These materials exhibit unusual behavior and show promise as potential low-temperature thermoelectric materials.

The electronic transport properties of transition-metal pentatellurides have intrigued scientists for more than two decades. A broad resistivity anomaly was found at a peak temperature T_P of 75 K in $HfTe_5$ [7.12] and 140 K in $ZrTe_5$ [7.13, 7.14] (Fig. 7.9). The resistivity versus temperature profile suggests that these materials exhibit metallic behavior below T_P and become semimetallic-like at higher temperatures. From solid solution studies of $Zr_{1-x}Hf_xTe_5$ by DiSalvo et al. [7.15] it was concluded that T_P varies systematically between transition-metal concentrations. To date, there remains no conclusive theory of the physics behind the peculiar temperature-dependent behavior of these systems. During the late 1970s and early 1980s, a great amount of interest in charge density wave (CDW) phenomena led to such investigations in the pentatelluride system; these phenomena are similar to the resistivity anomalies evident in $NbSe_3$. The consensus of these studies was that transition-metal pentatellurides did not exhibit CDW behavior due to the lack of nonlinearity in current–voltage curves, and to the absence of dual superlattice spots in diffraction patterns usually associated with these types of transitions [7.15, 7.16].

7.3.2 Crystal Structure and Band Structure

The isostructural parent compounds, $HfTe_5$ and $ZrTe_5$, form in the orthorhombic space group C*mcm* where the unit cell comprises four formula units. The unit cell dimensions are $a = 3.9743(5)$ Å, $b = 14.492(2)$ Å, $c = 13.730(2)$ Å; and $a = 3.9876(11)$ Å, $b = 14.502(4)$ Å, $c = 13.727(3)$ Å for $HfTe_5$ and $ZrTe_5$, respectively [7.17]. The transition-metal pentatellurides exhibit structure similar to that of trichalcogenides such as $NbSe_3$ and TaS_3, which are well known pseudo-one-dimensional conductors. The twenty-four-atom unit cell of the pentatellurides consists of four trigonal prismatic MTe_3 columns formed along the a axis interconnected by a pair of Te atoms parallel to the c axis and loosely coupled by

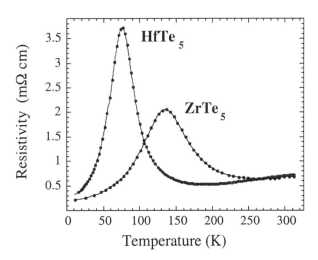

Fig. 7.9. Resistivity versus temperature for single crystals of $HfTe_5$ and $ZrTe_5$. Reprinted from [7.32]. Copyright 1999 American Physical Society

van der Waals forces along the *b* axis (Fig. 7.10). The resistivity measured along the *c* axis was two to three times greater than that along the *a* axis, a degree of anisotropy much lower than that of traditional low-dimensional conductors [7.16]. The anisotropy is only slightly temperature-dependent, which is also inconsistent with a CDW transition.

Single-crystal transition-metal pentatellurides are typically grown by an iodine transport technique [7.18]. Pentatellurides form shiny ribbon-like needles approximately 100 microns in diameter and a few millimeters long.

Fermi surface determinations of $ZrTe_5$ and $HfTe_5$ were interpreted using Shubnikov–de-Haas measurements [7.19–7.21]. As expected for an anisotropic orthorhombic material, the calculated Fermi surfaces were also anisotropic. $HfTe_5$ and $ZrTe_5$ were semimetals with three ellipsoidal pieces of Fermi surface, two electron and one hole. The $HfTe_5$ surfaces were more anisotropic and enclosed a volume an order of magnitude smaller than that of $ZrTe_5$.

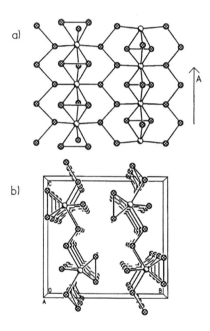

Fig. 7.10 (a) A unit cell of MTe_5 (M = Hf, Zr), along the *a* axis. The open spheres are the metal (M) atoms and the cross-hatched spheres are the Te atoms. (b) A projection of a layer in MTe_5 viewed down the *a* axis, showing the chains of MTe_3 pyramids and the van der Waals gap that separates the individual layers. Reprinted from [7.32]. Copyright 1999, American Physical Society

Band-structure calculations by Whangbo et al. [7.22] also predicted ellipsoidal Fermi surfaces. This study concluded that these materials were semimetals and that the conduction electrons around the Fermi level originate primarily from the Te atoms. This report also suggests that the resistivity anomaly would be less affected by substitutions at the transition-metal sites than those at the Te sites.

One very intriguing characteristic of these materials is their magnetoresistance. The magnetoresistance of transition-metal pentatellurides is very dependent on the orientation of the magnetic field [7.23, 7.24]. The largest magnetoresistance occurs when the magnetic field is parallel to the b axis ($B // b$), which is the axis that has the weakest bond interaction. The c axis is bridged between the metal chain prisms with Te bonds, which cross-link between these metallic chains. The effect of an applied magnetic field is highly anisotropic, and the magnetoresistance can vary substantially, depending on temperature and magnetic field strength. The magnetoresistance is low around room temperature, increases as the temperature is lowered, reaches a peak just below T_P, and undergoes a shallow minimum before increasing at lower temperatures. This is most apparent in the HfTe$_5$ material where $\rho(9T)/\rho(0) = 60$ at T_P. This behavior is similar to that observed in NbSe$_3$, although the magnetoresistance is much larger in HfTe$_5$.

7.3.3 HfTe$_5$ and ZrTe$_5$ Transport

As reported by Tritt et al. [7.25] and illustrated in Fig. 7.9, the temperature dependence of the resistivity for HfTe$_5$ exhibits a broad peak at around $T_P \approx 75$ K. At T_P, ρ is approximately six times that at room temperature. ZrTe$_5$ behaves similarly and has a higher temperature resistivity peak, $T_P \approx 140$ K, whose magnitude is three times larger than its room temperature ρ [7.14]. The room temperature ρ values are approximately 0.70 mΩ cm for HfTe$_5$ and 0.65 mΩ cm for ZrTe$_5$, which is comparable to the best Bi$_2$Te$_3$ alloys.

The α values of these materials also show behavior indicative of their potential for thermoelectric use (Fig. 7.11) [7.26, 7.27]. At high temperatures ($T >> T_P$), pentatellurides display a large positive (p-type) $\alpha > +125$ µV K^{-1}. Near T_P, α undergoes a dramatic change and passes through zero before reaching a negative (n-type) peak < -125 µV K^{-1}. Thus, these materials have α values that are relatively large over a broad temperature range and can have either sign, thereby exhibiting n-type ($T < T_P$) and p-type ($T > T_P$) behavior.

7.3.4 Isoelectronic Substitutions at the Transition-Metal Site

DiSalvo et al. [7.15] measured ρ and α of single crystals of Hf$_{1-x}$Zr$_x$Te$_5$ where x varied from $0 \leq x \leq 1$. T_P values determined by ρ versus temperature measurements shift to higher temperature as Hf is replaced by Zr. In each of the

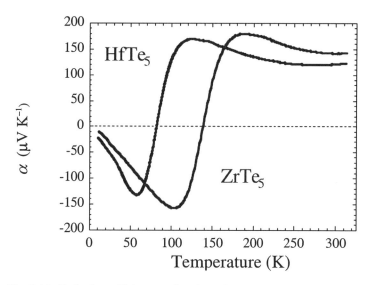

Fig. 7.11. Seebeck coefficient as a function of temperature for single crystals of HfTe$_5$ and ZrTe$_5$. Reprinted from [7.32]. Copyright 1999, American Physical Society

solid solution samples, ρ behaves similarly to that of the parent compounds with a systematic temperature shift. α of each sample also reveals a systematic shift in temperature as the Zr concentration is increased, similar to ρ, as shown in Fig. 7.12 [7.25]. Each concentration exhibits a relatively large p-type α at room temperature. As the temperature decreases, α in each sample increases until it reaches a maximum. At lower temperatures, α drops sharply, passes through zero at T_0, and continues to decrease until it reaches a maximum n-type α. At even lower temperatures, α begins to rise again toward zero in a relatively linear fashion as the temperature approaches absolute zero; this is characteristic of a diffusion mechanism. No apparent phonon-drag contribution was reported. As noted, the resistivity peak was first thought to be evidence of a charge density wave phenomenon; however, the origin of the anomaly in these pentatellurides, as discussed by DiSalvo et al. [7.15], appears to be an electronic phase transition as opposed to a structural phase transition. Therefore, the electronic properties of this system should be susceptible to doping which is evident from these results. The uncertainty in the concentrations of Hf or Zr does not allow predicting an obvious dependency of the $\alpha = 0$ temperature, T_0, or T_P. However, there is a distinct correlation between T_0 and T_P for each concentration.

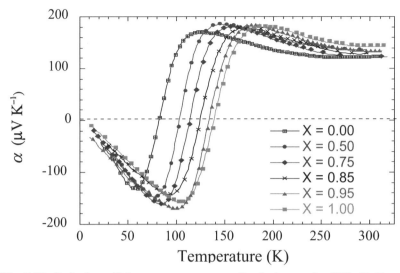

Fig. 7.12. Seebeck coefficient versus temperature for single crystals of $Hf_{1-x}Zr_xTe_5$

7.3.5 Titanium Substitution

The α values of pentatelluride materials are substantially affected by pressure. Below the peak, n-type α changes 150% or more to values of approximately -240 µV K^{-1} in $ZrTe_5$ at $T = 120$ K at a pressure of 12 kbar, and ρ decreases by a factor of 4 [7.28]. These trends enhance $\alpha^2\sigma$ by an order of magnitude. Smaller changes are observed in $HfTe_5$ under similar conditions. Uniaxial stress measurements have substantial effects on both parent materials. These results also favor the idea of an electronic phase transition.

Titanium atoms, whose metallic radius is 2.00 Å, are substantially smaller than either Hf or Zr; both have a radius of 2.16 Å, and therefore, should produce a slight compression of the lattice, possibly correlated to the applied external pressure. The Ti substitution is also isoelectronic (3d electrons instead of 4d or 5d) and hence should not directly alter the carrier concentration of the compounds. The ρ and α measurements are shown for $HfTe_5$ and $Hf_{0.95}Ti_{0.05}Te_5$ in Fig. 7.13 [7.29]. As reported by Littleton et al. [7.29], a small amount of Ti substitution (approximately 5%) shifts the peak temperature substantially from 75 K for $HfTe_5$ to $T_p \approx 40$ K for $Hf_{0.95}Ti_{0.05}Te_5$, but in contrast to $HfTe_5$, the zero crossing of α occurs at a much higher temperature ($T_0 \sim 50$ K) than T_p. The relative resistivity peak of the Ti-doped sample is nearly double that of the undoped resistivity peak. A high resistance state exists below the peak, compared to all other pentatelluride materials studied. This contrasts with the metallic behavior evident in other samples. The low-temperature, high-resistivity state is similar to that observed in $HfTe_5$

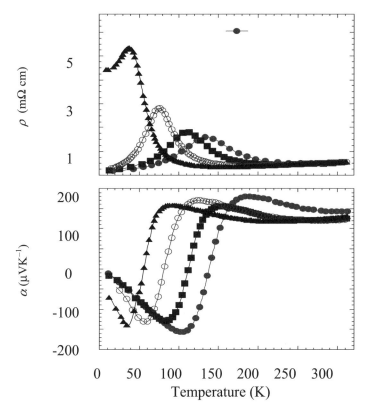

Fig. 7.13. Resistivity (**a**) and Seebeck coefficient (**b**) as a function of temperature for single crystals of $Zr_{0.90}Ti_{0.10}Te_5$ and $Hf_{0.95}Ti_{0.05}Te_5$ compared with $ZrTe_5$ and $HfTe_5$

under strain [7.30]. Figure 7.13 also shows similar ρ and α data for $ZrTe_5$ and $Zr_{0.90}Ti_{0.10}Te_5$. A nominal Ti substitution of 10% for Zr shifts the peak temperature from 140 K for $ZrTe_5$ to $T_p \approx 110$ K for $Zr_{0.90}Ti_{0.10}Te_5$, and T_0 coincides with T_p. The relative resistance peak is slightly larger than that of the parent $ZrTe_5$. The strong metallic behavior is again evident below T_p, and has a positive resistivity slope ($d\rho/dT > 0$). Substitutional doping of Ti for either Hf or Zr leads to a variation of the peak temperature from 38 to 140 K and maintains the relatively large values of α at low temperature.

7.3.6 Sb and Se Substitutions

Substitutional studies are important to find the combination of elements in a compound that produce certain desirable characteristics. Such studies performed on

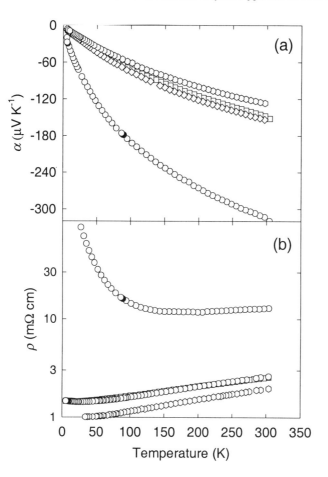

Fig. 6.19. Seebeck coefficient (**a**) and resistivity (**b**) of four n-type $Sr_8Ga_{16}Ge_{30}$ specimens with different Ga to Ge ratios but with similar Sr concentrations resulting in varying carrier concentrations. The room temperature Hall carrier concentration of these four specimens ranged from 10^{19} to 10^{20} cm^{-3}. Reprinted from [6.39]. Copyright 1998, American Institute of Physics

6.3.7 Summary and Conclusions

The clathrate structure, Ge clathrates, for example, can be thought of as a derivative of the four-coordinated diamond lattice structure of Ge. As described before, there is enough space between the Ge Ge atoms in the Ge diamond lattice to hold small interstitial atoms such as H or He. However, this space is not large enough to hold Sr or Eu atoms. The presence of these "guests" induces a change to the more open Ge clathrate structure.

Bi$_2$Te$_3$ led to a series of related alloys and pseudoternary compounds that resulted in effectively optimized thermoelectric materials [7.31, 7.32]. This optimizing process found that specific stoichiometric amounts of Se and Sb enhanced the Bi$_2$Te$_3$ compounds. Antimony is an element in many thermoelectric materials, including Bi$_2$Te$_3$ and of course Bi–Sb alloys. Bismuth–antimony alloys have extremely large magnetothermoelectric figure of merit values below room temperature. Littleton et al. [7.32] showed that small amounts of antimony-doped pentatellurides, shown in Fig. 7.14, did not improve the thermoelectric properties at lower temperatures. Instead, the Sb completely alters the temperature dependency of both α and ρ. The 5% nominally doped pentatellurides

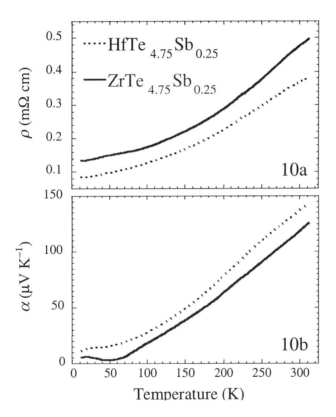

Fig. 7.14. Resistivity (**a**) and Seebeck coefficient (**b**) as a function of temperature for single crystals of HfTe$_{4.75}$Sb$_{0.25}$ and ZrTe$_{4.75}$Sb$_{0.25}$

HfTe$_{4.75}$Sb$_{0.25}$ and ZrTe$_{4.75}$Sb$_{0.25}$ both exhibit α and ρ values that increase with temperature. The samples measured had relatively low room temperature ρ values of approximately 0.48 and 0.36 mΩ cm for HfTe$_{4.75}$Sb$_{0.25}$ and ZrTe$_{4.75}$Sb$_{0.25}$, respectively. The HfTe$_{4.75}$Sb$_{0.25}$ samples have temperature-dependent values that are larger in α and lower in ρ than those of ZrTe$_{4.75}$Sb$_{0.25}$. The autonomous transition, which occurs in the parent materials, apparently, no longer exists within these doped materials. The possibility of a dissimilar structural phase was eliminated by X-ray analysis that reconfirmed the pentatelluride structure.

Results from Se substitutions at Te sites of both parent pentatelluride materials are quite promising as well [7.32]. As seen in Fig. 7.15, a 5% nominal substitution of Se for Te reduces the temperature dependency of T_p and T_0. T_p and T_0 are reduced nearly 20 K from HfTe$_5$ to HfTe$_{4.75}$Se$_{0.25}$, and reduction is only a few kelvin for ZrTe$_5$ to ZrTe$_{4.75}$Se$_{0.25}$. The magnitudes of the absolute α for both Se-doped materials increase approximately 20% from those of their parent materials. α values exceeding 200 μV K^{-1} were measured for both HfTe$_{4.75}$Se$_{0.25}$ and ZrTe$_{4.75}$Se$_{0.25}$ at temperatures of approximately 95 and 185 K, respectively. Another favorable effect of the Se substitution is the reduction in ρ. The ρ values

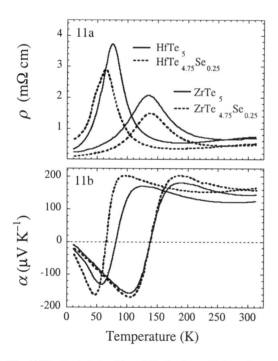

Fig. 7.15. Resistivity (**a**) and Seebeck coefficient (**b**) versus temperature for single crystals of MTe$_{5-x}$Se$_x$ (M=Hf, Zr, and x = 0, 0.25). Reprinted from [7.32]. Copyright 1999, American Physical Society

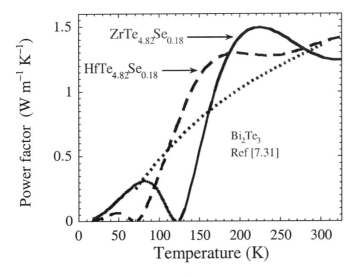

Fig. 7.16. Power factor $\alpha^2 T/\rho$ (or $\alpha^2 \sigma T$) as a function of temperature for HfTe$_{4.82}$Se$_{0.18}$ and ZrTe$_{4.82}$Se$_{0.18}$, compared to optimized Bi$_2$Te$_3$. Reprinted from [7.32]. Copyright 1999, American Physical Society

of the parent pentatellurides decrease approximately 25% with the nominal addition of 5% Se for Te. The increase in α, combined with the decrease in ρ, results in an enhancement of $\alpha^2 T/\rho$ by a factor of 2. $\alpha^2 T/\rho$ values of Se-doped pentatellurides range from ≥ 1.25 W m^{-1} K^{-1} at 150 K $\leq T \leq$ 320 K for Hf(Te$_{1-x}$Se$_x$)$_5$ to ≤ 1.50 W m^{-1} K^{-1} at 225 K $\leq T \leq$ 320K for Zr(Te$_{1-x}$Se$_x$)$_5$. The significance of these results is illustrated in Fig. 7.16, where the $\alpha^2 T/\rho$ of Se-doped samples are compared to that of an optimally doped Bi$_2$Te$_3$ alloy by Yim and Rosi [7.31]. At low temperatures, 150 K $< T <$ 250 K, the power factor of pentatellurides significantly exceeds that of the state-of-the-art bismuth telluride. Thus far, the small size of these single crystal materials has made accurate ($\approx$ 10%) determinations of λ very difficult. Nevertheless, these results for the electronic properties are very encouraging.

7.4 Tl$_2$SnTe$_5$ and Tl$_2$GeTe$_5$

The Tl–Sn–Te [7.33] and Tl–Ge–Te [7.34] systems both contain several ternary compounds, including a 2–1–5 composition in both systems. Some crystal structure data for Tl$_2$SnTe$_5$ and Tl$_2$GeTe$_5$ are given in Table 7.1. Both compounds are tetragonal and contain columns of Tl ions along the crystallographic c axis. The

Table 7.1. Crystal data for Tl$_2$SnTe$_5$ and Tl$_2$GeTe$_5$. In each case, the second ADP value is from neutron diffraction data

	Tl$_2$SnTe$_5$[a] Tetragonal x-ray density = 7.40 g cm^{-3}	
a = 8.306 Å	Tl–Te distances (Å)	ADPs (Å^2)
c = 15.161 Å		
I4/mcm	Tl(1): 8 Te at 3.49	Tl(1): 0.017, 0.026
32 atoms cell^{-1}	Tl(2): 8 Te at 3.66	Tl(2): 0.047, 0.049
	Tl$_2$GeTe$_5$[b,c] tetragonal X-ray density = 7.34 g cm^{-3}	
a = 8.243 Å	Tl–Te distances (Å)	Isotropic thermal parameters
c = 14.918 Å		
P4/mbm	Tl(1): 4 at 3.53, 4 at 3.79	Tl(1): 0.048, 0.042
32 atoms cell^{-1}	Tl(2): 4 at 3.38, 4 at 3.54	Tl(2): 0.025, 0.021

[a] Ref. 7.35
[b] Ref. 7.34
[c] Ref. 7.40

large interatomic distances for the eightfold-coordinated Tl ions (Table 7.1) are one of the main reasons that these compounds were selected as thermoelectric candidates. Transverse to the c axis, these columns alternate with chains of composition (Sn/Ge)Te$_5$. In Tl$_2$SnTe$_5$ (Fig. 7.17), the chains can be described as SnTe$_4$ tetrahedra linked by Te atoms that are in square-planar coordination. The GeTe$_5$ chains in Tl$_2$GeTe$_5$ are better described as the alternation of Te$_4$ square rings and edge-sharing pairs of GeTe$_4$ tetrahedra (corresponding to a composition of Ge$_2$Te$_6$ for each pair of tetrahedra). Tl$_2$SnTe$_5$ and Tl$_2$GeTe$_5$ can be viewed as polytypes of one another, where different stacking sequences of (Ge/Sn)Te$_4$ tetrahedra and TeTe$_4$ square planar units are linked into chains by edge sharing [7.35].

In the course of studying the thermoelectric properties of Tl$_2$SnTe$_5$ and Tl$_2$GeTe$_5$, Sharp et al. [7.36] performed neutron diffraction on powders to refine the structure and calculate ADPs. Large ADP values confirm that a portion of the Tl atoms are loosely bound in these structures. Again large ADP values are associated with abnormally low values of λ_L, [7.37] as discussed later.

Polycrystalline samples (~ 6 x 6 x 13 mm^3) for transport measurements [7.36] were taken from quenched and annealed specimens or from hot-pressed specimens. Small single crystals (~ 2.5 x 2.5 x <1 mm^3) were grown (flux composition approximately TlTe$_2$) and used for electrical measurements in the ab plane. Both Tl$_2$SnTe$_5$ and Tl$_2$GeTe$_5$ are stable in air, water, and common solvents. The transport properties of four samples are shown in Figs. 7.18–7.20.

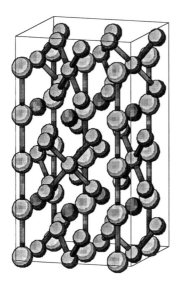

Fig. 7.17. Crystal structure of Tl_2SnTe_5. The larger spheres represent Tl, and the darker small spheres represent Sn. There are no Tl–Tl bonds, but the Tl atoms have been connected by thin lines to show their arrangement along the (vertical) c axis. $TeTe_4$ square planar units have been emphasized. For clarity, the Sn–Te (tetrahedral Sn) and Tl–Te bonds are not shown. There are two eightfold sites for Tl: cubic and square antiprism. Reprinted from [7.36]. Copyright, American Institute of Physics

The α curves (Fig. 7.18) indicate that Tl_2SnTe_5 and Tl_2GeTe_5 are small band-gap semiconductors and that undoped samples are p-type. From the maxima of α as a function of temperature [7.10], we estimate the band gaps at approximately 0.25 and 0.17 eV for Tl_2SnTe_5 and Tl_2GeTe_5, respectively. Hall measurements on a Tl_2SnTe_5 sample with $\alpha \cong 210$ $\mu V\ K^{-1}$ at room temperature yielded a carrier concentration (n_p) and mobility (μ_p) of $n_p = 2.8 \times 10^{19}$ cm^{-3} and $\mu_p = 55$ cm^2 V^{-1} s^{-1}. Similarly, for a Tl_2GeTe_5 sample with $\alpha = 270$ $\mu V\ K^{-1}$ at 300 K, Sharp et al. found $n_p = 2.0 \times 10^{19}$ cm^{-3} and a mobility $\mu = 26$ cm^2 V^{-1} s^{-1}.

The behavior of ρ as a function of temperature (Fig. 7.19) is not straightforward. In the initial work, some quenched/annealed or hot-pressed samples exhibited a resistivity that decreased approximately linearly with T between 150 and 300 K. However, the resistivity of other quenched/annealed or hot-pressed samples reached a minimum between 200 and 250 K and increased as the temperature was lowered further. This behavior might indicate a contribution from cracking and/or grain-boundary resistance. Further, the resistivity of some samples displayed hysteresis as the temperature was cycled and was lower during cooling than during warming.

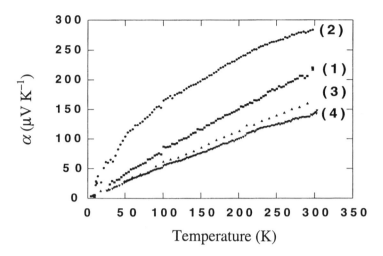

Fig. 7.18. Seebeck coefficient data for three polycrystalline samples and one single-crystal sample: (1) Tl_2SnTe_5, (2) Tl_2GeTe_5, (3) hot-pressed Tl_2SnTe_5, (4) Tl_2SnTe_5 single crystal. Samples 1 and 2 were melted, quenched, and annealed. Electrical properties were measured in the *ab* plane of the platelet-shaped single crystal. Reprinted from [7.36]. Copyright, American Institute of Physics

The thermal conductivity is very low for both Tl_2SnTe_5 and Tl_2GeTe_5 (Fig. 7.20). Apparently λ_L in these polycrystalline samples is not more than 5 mW cm^{-1} K^{-1}. This is less than one-third of the value for that of pure Bi_2Te_3. At room temperature, the electronic contribution to λ is estimated at about 20% for Tl_2SnTe_5, but less than 10% for Tl_2GeTe_5 due to the higher resistivity.

The best results from the initial work on this compound (sample 1 in Figs. 7.18, 7.19, and 7.20), give a zT of 0.6 at 300 K and an estimated peak of 0.85 at 400 K. For comparison, the maximum zT for Bi_2Te_3 is 0.6 at 300 K. The practical value of Tl_2SnTe_5 and Tl_2GeTe_5 is limited by the extreme toxicity of the oxides of Tl, but other compounds with this structure, such as K_2SnTe_5 [7.38] and Rb_2SnTe_5 [7.39], may also have good thermoelectric properties.

7.4 Tl$_2$SnTe$_5$ and Tl$_2$GeTe$_5$ 233

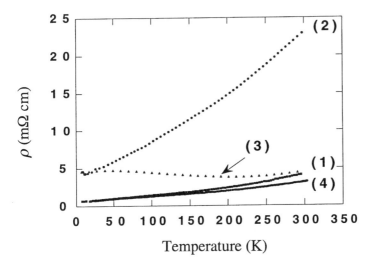

Fig. 7.19. Electrical resistivity data for three polycrystalline samples and one single-crystal sample. The geometry (length-to-area ratio) uncertainty is ±15%. Refer to Fig. 7.18 for sample descriptions. Reprinted from [7.36]. Copyright, American Institute of Physics

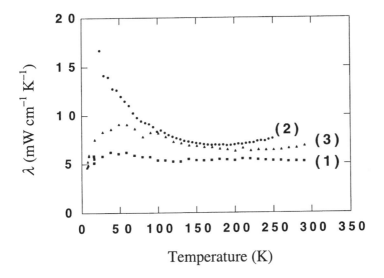

Fig. 7.20. Thermal conductivity data for three polycrystalline samples. The geometry (length-to-area ratio) uncertainty is ±15%. Refer to Fig. 7.18 for sample descriptions. Reprinted from [7.36]. Copyright, American Institute of Physics

8. Low-Dimensional Thermoelectric Materials

In Chap. 2, the thermoelectric transport coefficients (Seebeck, α, resistivity, ρ, and thermal conductivity, λ) were defined and analyzed as if they were strictly bulk quantities. This assumption is not valid if boundaries significantly affect the transport of electric current or heat, which can result from overall sample dimensions or structural features. The topic of size effects has become important in thermoelectric materials research because theoretical and experimental enhancements of the dimensionless figure of merit zT have been reported for low-dimensional structures. This area of the field is developing rapidly, and most of the work mentioned here has been done since 1990. First, we discuss, though, a few important earlier developments that comprise a useful prologue to the more recent work that we outline here.

8.1 Fine-Grained Si–Ge and Thin-Film Bi

The dependence of the lattice thermal conductivity λ_L on sample size was observed and explained as long ago as 1938 [8.1]. It was realized that the effect was strongest at low temperatures because the phonon mean free path due to Umklapp scattering became comparable to, or even greater than, the dimensions of crystalline samples. At higher temperatures, Umklapp scattering reduces the mean free path to values far below the size of typical bulk samples.

Goldsmid and Penn [8.2] pointed out that it is instructive to distinguish the contributions to λ_L from low-, moderate-, and high-frequency phonons. Because point defects scatter mainly high-frequency phonons and Umklapp scattering is likely to dominate in the middle of the spectrum, heat conduction in solid solutions at ordinary temperatures is skewed toward low-frequency phonons. Thus boundary scattering, which impacts mainly low-frequency phonons, is complementary to alloy and Umklapp scattering and relevant to thermoelectric materials. A simple analysis shows that boundary scattering can be surprisingly effective in fine-grained solid solutions [8.3], particularly Si–Ge alloys [8.4], that occurs at higher temperatures and greater grain size-to-mean free path ratios than one might expect at first. Measurements of λ_L of $Si_{70}Ge_{30}$ as a function of grain size have confirmed the theoretical estimates [8.5].

In the theory of boundary scattering of phonons in alloys just mentioned, one assumes a phonon spectrum that is unaffected by the boundaries. One also assumes that the scattering is random. The influence of the boundaries is simply to limit the free path of those phonons that would otherwise have a long free path. Such scattering can be termed a "classical size effect" because it does not take into account the restructuring of momentum space due to size effects. "Quantum size effects" occur when the alteration of momentum space cannot be ignored. The distinction between classical and quantum size effects was invoked in the study of ρ [8.6] and α [8.7] of thin Bi films.

The dependence of α on film thickness is due to the fact that boundary scattering of electrons may have an energy dependence different from bulk processes. In the classical approximation for metals, this leads to an additional term in α that depends on the ratio λ_0/d, where λ_0 is the bulk mean free path of electrons and d is the film thickness. In a semimetal such as Bi, one might expect that the situation could be complicated if boundary scattering changes the balance of partial α's of electrons and holes. The key point, however, is that a transition will be seen as the film thickness drops below the de Broglie wavelength λ_{dB} of the carriers [8.7]. For $d < \lambda_{dB}$, the quantization of momentum space becomes important.

These historical examples introduce an important duality in the study of size effects in thermoelectric materials: both the mean free paths and the wavelengths of electrons and phonons must be considered in a general analysis.

8.2 Survey of Size Effects

The power factor $\alpha^2\sigma$ depends on the scattering parameter, the density of states, the mobility, and the position of the Fermi level. Historically, the first three have usually been treated as intrinsic material properties that can be improved only by making better samples (purer, denser, etc.). Only the Fermi level has been used routinely as a tool for optimizing $\alpha^2\sigma$. Similarly, λ_L depends on the specific heat, the sound velocity, and the mean free path of phonons. Solid solutions and, occasionally, grain size studies have been used to minimize the mean free path, but there has been no means to reduce sound velocity. By contrast, artificially structured materials offer opportunities to engineer a broad range of material properties, including those pertinent to raising $\alpha^2\sigma$ and lowering λ_L.

8.2.1 Electron Confinement

We saw in Chap. 3 that z increases with the parameter β, which is proportional to the density of states that lie within a few $k_B T$ of the Fermi level. For this reason, a large effective mass is beneficial if it is not offset by an equally large inertial mass, which appears in the denominator of the mobility. Certain anisotropic semiconductors provide such a beneficial ratio of effective mass to inertial mass, provided that transport is arranged in the preferred direction. As Hicks and Dresselhaus [8.8] pointed out, this situation can be mimicked and exaggerated in two-dimensional (2-D) semiconductor systems.

Consider a single layer of a small bandgap semiconductor (the well) sandwiched between two layers of semiconductor with a larger band gap (the barriers), where the interfaces are taken as ideal. Impurity levels are adjusted so that the well is heavily n-type and the barriers are lightly doped. Electrical conduction will take place mainly in the well (x–y plane), and the conductivity along the x axis in a 2-D approximation is given by

$$\sigma_{2D} = \frac{1}{2\pi a}\left(\frac{2k_B T}{\hbar^2}\right)\left(\frac{m_y}{m_x}\right)^{1/2} F_0 e^2 \tau_0, \tag{8.1}$$

where a is the well thickness, which is in the z direction. We have assumed that only the lowest subband is occupied, that the electron relaxation time is independent of energy, and that the transport is in the x direction. For comparison to the 3-D situation, Eq. (2.54) generalized for an anisotropic electron mass and with $r = 0$ is

$$\sigma_{3D} = \frac{1}{2\pi^2}\left(\frac{2k_B T}{\hbar^2}\right)^{3/2}\left(\frac{m_y m_z}{m_x}\right)^{1/2} F_{1/2} e^2 \tau_0. \tag{8.2}$$

In the Fermi–Dirac integrals $F_n(\xi)$ as defined in (2.53), the Fermi level ξ is measured from the respective band edge, which is raised by $\hbar^2/8m_z a^2$ in the 2-D case.

With the same assumptions as given for 2-D conductivity, the 2-D Seebeck coefficient is

$$\alpha_{2D} = -\frac{k_B}{e}\left(\frac{2F_1}{F_0} - \xi\right), \tag{8.3}$$

and the corresponding 3-D result (2.55) is

$$\alpha_{3D} = -\frac{k_B}{e}\left(\frac{5F_{3/2}}{3F_{1/2}} - \xi\right). \tag{8.4}$$

The ratio α_{2D}/α_{3D} is only a weak function of ξ, and $(\alpha_{2D}/\alpha_{3D})^2 \approx 2/3$ for $-1 < \xi < 1$. For a given ξ, there is no 2-D enhancement of α. The ratio σ_{2D}/σ_{3D} has also a rather weak dependence on ξ in the vicinity of $\xi = 0$, the optimum range for thermoelectric materials. For any nearly optimal ξ, we find that $\sigma_{2D}/\sigma_{3D} \approx h/2(2m_z k_B T)^{1/2} a$. This can be rewritten as $\sigma_{2D}/\sigma_{3D} \approx \lambda_{dB}/2^{1/2} a$ where λ_{dB} is the de Broglie wavelength for an electron with energy of $2k_B T$ and mass m_z. It is clear that the well width a must be less than λ_{dB} to enhance electrical properties by quantum confinement. Numerically, at 300K, $\sigma_{2D}/\sigma_{3D} \approx 3.86$ nm/$(m_z)^{1/2} a$, where m_z is now in units of free electron mass. Thus, there are general conditions for enhancing $\alpha^2 \sigma$ in 2-D compared to 3-D, provided that τ_0 does not change. For $r = 0$, the condition is that $(m_z)^{1/2} a < 2.5$ nm. For $m_z = 0.1$, $a < 8$ nm is required.

The assumption that the barrier layers completely isolate a well from adjacent wells requires a large barrier height or barrier thickness. A thick barrier will cause a thermal short of the well if the barrier and well have a similar λ_L. Therefore, large barrier heights are essential because this allows the barrier thickness to be small. Broido and Reinecke [8.9] have shown that a zT increase of 50% (at a repeat distance of 5 nm) might be achieved for Bi_2Te_3 quantum well superlattices with 1 eV barriers.

The theoretical benefit of quantum confinement also applies to quantum wires, that is one-dimensional (1-D) conductors [8.10]. For an advantageous effective mass tensor, the possibility exists for large enhancements of zT_{1-D} in wire diameters less than 5 nm. However, if similar barrier materials are available as for quantum wells, then parasitic heat conduction is more detrimental because the wires will constitute a smaller fraction of the sample volume [8.11]. Therefore, large energy barriers are still more important for 1-D thermoelectric materials.

8.2.2. Band Engineering

The band structure of layered semiconductors may differ in important ways from the corresponding bulk materials. The simplest difference that can impact zT is the increase of the band gap that often occurs in quantum wells. For instance, quantum confinement increases the magnitude of the band gap in Bi [8.12], which then becomes a more extrinsic conductor with less degradation from simultaneous electron and hole conduction. This is one reason that Bi nanowires are being studied experimentally, as discussed in Sect. 8.3.5.

Other, more subtle effects may yield quite large increases in zT. The relative positions of different band extrema will change in a superlattice. These relative changes can be caused by confinement (the energy shift of each extremum depends on the associated carrier effective mass) or by strain (each extremum has its own strain deformation potential). Thus, by varying the strain-free lattice constant (i.e., the composition) of the substrate and epitaxial layers and the thickness of the layers, it is possible to maximize zT. Specifically, the two goals are to enhance band offsets that provide confinement and to maximize the density of states by aligning extrema. Either of these effects could dramatically enhance $\alpha^2\sigma$ relative to the bulk material, and achieving both simultaneously can lead to very large zT increases. In one instance, calculations indicate nearly a 70-fold increase, relative to bulk Si, for the room temperature zT of a Si(1.5 nm)/Ge(2.0 nm) superlattice grown on (111) $Si_{0.5}Ge_{0.5}$ [8.13]. In this system, the L valleys in Ge and Δ valleys in Si contribute to transport, and both sets of valleys experience some degree of confinement.

8.2.3 Boundary Scattering of Electrons

Another possibility is that, for a given carrier concentration, σ and α could change as a result of a change in the effective scattering parameter [8.14]. Suppose that the bulk scattering mechanism results in a lifetime of τ_1 and $\alpha = \alpha_1$. Imagine, then, that a second scattering mechanism due to size effects is introduced with associated parameters of τ_2 and α_2. The conductivity will decrease by the factor $\sigma/\sigma_1 = \tau_2/(\tau_1 + \tau_2)$, and

$$\alpha/\alpha_1 = \tau_2/(\tau_1 + \tau_2) + (\alpha_2/\alpha_1)\tau_1/(\tau_1 + \tau_2). \tag{8.5}$$

With these assumptions, $(\alpha/\alpha_1)^2(\sigma/\sigma_1)$ can be noticeably greater than unity, provided that $\alpha_2/\alpha_1 > 1.6$. If $\alpha_2/\alpha_1 = 2$, then nearly a 20% increase in the power factor is possible for $\tau_2 \cong 2\tau_1$. Although the Lorenz number will increase in this example, the decrease of the conductivity will result in little change in the electronic component of the thermal conductivity λ_e, so there is hope that zT could be increased.

8.2.4 Modulated Umklapp Scattering

Ren and Dow [8.15] introduced the concept of mini-Umklapp scattering in superlattices. In mini-Umklapp scattering, a three-phonon event, the requirement is that the initial and final wave vectors differ by a minireciprocal lattice vector (Fig. 8.1). If two materials with lattice constant a and mass difference ΔM are formed in an $N \times N$ superlattice, then there are Umklapp scattering events associated with each of the $2N-1$ minireciprocal lattice vectors. The energy of these phonons is less than that of the phonons required for Umklapp scattering in the bulk materials, and so low-temperature Umklapp scattering is enhanced. The strength of such scattering is expected to be proportional to $(\Delta M)^2$ and inversely proportional to N^2. The decrease in λ_L at the low temperature peak is marginal (< 25%), rather than factorial.

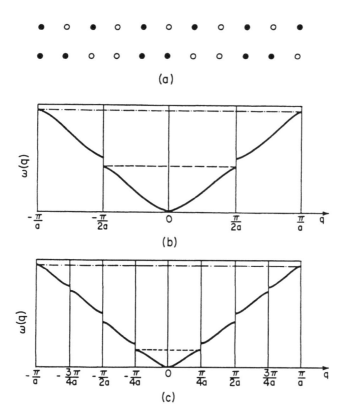

Fig. 8.1. Illustrations of 1x1 and 2 x 2 superlattices (a) and their respective Brillouin zones (b), (c). The 1 x 1 superlattice has a bulk Umklapp process wave vector (dashed-dotted line) and a mini-Umklapp process wave vector (dashed line). The 2 x 2 superlattice has a bulk process wave vector (dashed-dotted line) and three mini-Umklapp process wave vectors, the smallest is denoted by a dashed line. Reprinted from [8.15]. Copyright 1981, American Physical Society

8.2.5 Boundary Scattering of Phonons

The reduction of λ_L in superlattices is more than an order of magnitude for systems such as Si/Ge. Mini-Umklapp scattering cannot explain an effect this strong. Another contribution to the decrease of λ_L seems to be due to boundary scattering of phonons, which are treated as classical waves. Chen and Neagu [8.16] modeled both specular and diffuse scattering. The strength of diffuse scattering, allowing inelastic events, depends on the mismatch of the product Cv, where C is the volumetric specific heat and v is the phonon group velocity. The strength of (elastic) specular scattering depends on the mismatch of the product ρv, where ρ is the mass density, and on the angle of incidence. For specular scattering, there is a critical angle above which total internal reflection of phonons occurs. Phonons on one side of a specular interface can also be confined for any angle of incidence if their frequency is above the maximum acoustic phonon frequency in the other material, provided that inelastic scattering does not occur [8.17].

As one would expect, diffuse scattering gives a much stronger reduction of λ_L than specular scattering. This is especially true for in-plane thermal transport. For diffuse scattering, the cross-plane λ_L of a Si–Ge superlattice falls below that of a $Si_{0.5}Ge_{0.5}$ alloy for periods less than ~ 10 nm, this gives a result consistent with experimental data [8.18] for periods between 3 and 7 nm. Relative to the average of the bulk λ_L's, there is a 30-fold reduction in these short period superlattices. For AlAs–GaAs superlattices, the reduction of in-plane λ_L measured by Yao [8.19] is not nearly as great as for Si–Ge, a fact that is attributed to a smaller mismatch of acoustic properties [8.16].

Boundary scattering of phonons may also impact α. If there is a significant phonon-drag effect, this part of α is sensitive to any phonon scattering mechanism. Further, the phonon-drag contribution is associated mainly with long wavelength phonons and thus may be particularly sensitive to boundary scattering. Millimeter-sized Ge crystals at low temperatures have been used for experimental studies of the effect of boundary scattering of phonons on both λ_L [8.20] and α [8.21].

8.2.6 Phonon Dispersion

Another approach to the superlattice λ_L problem is to emphasize the modification of the phonon dispersion curves relative to the bulk dispersion curves of the constituent layers. The ratio λ_L/τ_p, where τ_p is the (frequency-independent) phonon relaxation time, can be calculated from the Boltzmann transport equation and isolates the contribution of reduced phonon group velocity in superlattices [8.22]. For a superlattice with a period composed of two layers each of Si and Ge, λ_L/τ_p at ordinary temperatures for cross-plane transport is approximately an order of magnitude less than the average of λ_L/τ_p for Si and Ge. The reason is that the average phonon group velocity is greatly reduced. In fact, there are two portions of the phonon spectra for which the group velocity is effectively zero. First, the high-frequency portion of the Si spectrum contains modes that are not allowed in Ge. Second, a fraction of the phonons in Ge corresponds to an exponentially decaying wave in Si, which is total internal reflection described in terms of equilibrium phonon distributions. Neither group of confined phonons will contribute to heat conduction perpendicular to the interfaces. At low temperatures the average group velocity for the superlattice is an average of the corresponding Si/Ge velocities, and there is no reduction of λ_L/τ_p.

8.2.7 Boundary Physics

Finally, it may be that a satisfactory description of low-dimensional thermoelectric materials and devices will require treating the impact of boundary impedances on electron–phonon equilibrium [8.23]. In general, we may expect that there are boundary corrections to the flow of heat and electricity that become more important as the distances between boundaries become smaller. The disruption of electron–phonon equilibrium at the boundary will give way to bulk equilibrium over some characteristic distance. If the electrical and thermal gradients within the out-of-equilibrium region are a significant part of the total gradients, then an accurate description must include this additional complication. It has been found, for instance, that the type of electron conduction across the boundary affects the numerical factor in the boundary Wiedemann–Franz law. Models of emission over a barrier and tunneling through a barrier give Lorenz numbers of 2 and $\pi^2/3$, respectively, matching the nondegenerate and degenerate limits of bulk λ_e.

8.3 Experimental Structures

During the last three decades, engineered structures that have submicron dimensions have become prevalent in physics and materials science. As an alternative to seeking new materials, it is possible to change the properties of known materials by exploiting size effects. In the field of thermoelectric materials, both 2-D and 1-D structures are being explored in the search for higher zT values.

8.3.1 Quantum Wells

Whall and Parker [8.24] gave the first discussion of quantum well transport in the context of thermoelectric materials, and suggested that enhanced properties could be observed in Si–Ge multilayers grown by molecular beam epitaxy (MBE). These authors noted that, according to experimental [8.25] and theoretical [8.26] results, a 2-D electron gas is formed in the Si layers of $Si/Si_{0.5}Ge_{0.5}$ superlattices grown on $Si_{0.75}Ge_{0.25}$ buffers on Si substrates. The localization of the electron gas results from an ~ 0.15 eV conduction band offset that is almost entirely due to strain effects. In the same system, a 2-D hole gas can be formed in the $Si_{0.5}Ge_{0.5}$ layers, regardless of the strain state.

Following the conjectures of Whall and Parker [8.24], MBE has proven to be a useful tool for demonstrating the enhancement of thermoelectric properties in quantum wells. Harman and co-workers [8.27] synthesized and measured the electrical properties of n-type PbTe/Pb$_{1-x}$Eu$_x$Te superlattices (typically >100 periods) grown on BaF$_2$ substrates after depositing a buffer layer (Fig. 8.2). In this system, the PbTe layers are the wells, and the conduction band of Pb$_{0.93}$Eu$_{0.07}$Te is approximately 0.2 eV higher in energy. Bi, a donor, was incorporated into the barrier layers. Although the barrier layer thickness was in the range of 40 to 60 nm, calculations show that the 2- to 5-nm PbTe layers dominate electrical conduction. Measurements show that for a given electron concentration α was enhanced by nearly a factor of 2 for 2-nm wells. No increase was seen for 4-nm wells. The electron mobility in these multiquantum well structures ranged from 40 to 90% of the mobility in good bulk samples. Even with degraded mobilities, there was a considerable increase of $\alpha^2\sigma$ for the 2-nm wells that approached a threefold increase for their best samples. In the light of the decreased σ, no increase in λ_e is expected, and it is safe to assume that λ_L of the PbTe wells is not greater than that of bulk PbTe. Thus, the zT of the well is enhanced by at least as much as the power factor, and the 2-D zT is estimated to be as high as 1.2 at 300 K. In these samples, however, the large ratio of barrier volume to well volume causes a low overall zT.

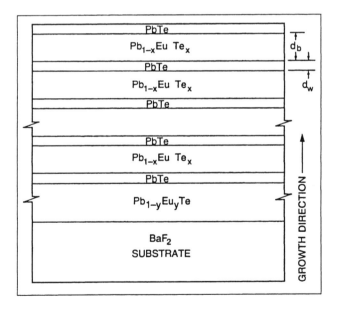

Fig. 8.2. Schematic cross-section of PbTe/Pb$_{1-x}$Eu$_x$Te multiple quantum wells in which the 2-D enhancement of zT was demonstrated. Reprinted from [8.27] with permission

8.3.2 Nonquantum Well Superlattices

The properties of quantum wells are measured in-plane. A different approach is to create nonquantum well structures and focus on cross-plane transport. In this scheme, each constituent should have good electrical properties because the current will flow through all layers. The goal is to achieve a large reduction of λ_L but leave the electrical properties intact.

Two groups have reported cross-plane transport measurements for Bi_2Te_3/Sb_2Te_3 superlattices. Excellent quality Bi_2Te_3/Sb_2Te_3 structures can be grown on [100] GaAs by modified organometallic chemical vapor deposition (CVD), as evidenced by TEM (Fig. 8.3). It is thought that the van der Waals gaps of the crystal structure allow virtually defect-free layers to form, even in cases of large lattice constant mismatch with the substrate. For these high-quality layers, the in-plane hole mobilities exceed those of the equivalent bulk alloys, even for the minimum 2-nm periods [8.28]. Yamasaki et al. [8.29] showed that smooth Bi_2Te_3/Sb_2Te_3 superlattices can be synthesized by pulsed laser ablation of Bi_2Te_3 and Sb_2Te_3 targets. Lower λ was observed for superlattices than for films, but no electrical properties were reported. For the CVD superlattices, the

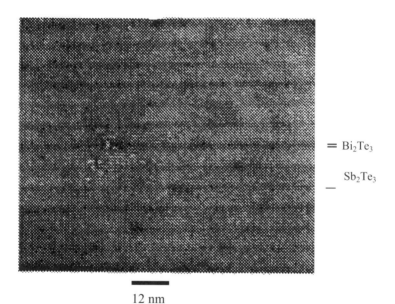

12 nm

Fig. 8.3. Transmission electron microscope image of a 10 Å/50 Å Bi_2Te_3/Sb_2Te_3 superlattice grown by MOCVD on GaAs that showed very sharp interfaces. Ten angstroms corresponds to the minimum repeat distance, the thickness of one Te–Bi–Te–Bi–Te packet. Reprinted from [8.28]. Copyright 1999, American Institute of Physics

cross-plane λ measured by the 3ω method is lower than that of equivalent alloys for all periods in the range of 2 to 20 nm [8.30]. Values as low as 0.2 W m^{-1} K^{-1} are obtained for structures in which the Bi$_2$Te$_3$ and/or Sb$_2$Te$_3$ layers are 3 nm thick. Likewise for the pulsed laser ablation structures, the cross-plane λ measured by an ac calorimetric method is less than that of bulk alloys for all periods in the range of 6 to 80 nm. The minimum λ in this case was 0.11 W m^{-1} K^{-1} for equal Bi$_2$Te$_3$ and Sb$_2$Te$_3$ layer thicknesses of 6 nm. For both the CVD and PLA films, it appears that the data presented are total λ, in which case λ_L must be lower yet. The zT for these superlattices is not available because no cross-plane electrical measurements were made.

Si–Ge is another good bulk thermoelectric materials system that has been studied in superlattice form. Si–Ge superlattices have been synthesized by CVD on GaAs substrates [8.31] and by MBE on Si substrates after depositing a graded Si$_{1-x}$Ge$_x$ buffer layer [8.32]. In CVD-grown layers, the measured λ reached a minimum of 2 W m^{-1} K^{-1} for a 1.4 μm thick superlattice with a period composed of 10-nm Si and 4-nm Ge. This value is only 50% greater than amorphous Si, and the temperature dependence was glass-like. Much of the thermal resistance was attributed to a high density of dislocations and stacking faults. Interestingly, superlattices that had thinner constituent layers had higher λ [8.31] (Fig. 8.4). In MBE-grown layers, a symmetrical 7.3-nm period superlattice 1.1 μm thick was tested. A differential 3ω technique was used to resolve the in-plane and cross-plane λ's. The results were $\lambda = 3.4$ W m^{-1} K^{-1} cross-plane and 1.4 W m^{-1} K^{-1} in-plane. The lower in-plane value is unexpected and could be due to threading dislocations or to measurement errors, which were larger for the in-plane direction. The cross-plane λ is two to three times lower than that of a bulk Si$_{0.5}$Ge$_{0.5}$ alloy [8.32]. These Si–Ge superlattice λ data provide a unique opportunity to compare results from different groups studying similar low-dimensional thermoelectric structures. For a nearly symmetric superlattice with a period of 6.5 nm, comparable to the 7.3 nm period structure described before, a cross-plane λ of ~ 4.5 W m^{-1} K^{-1} [8.31] was reported. The uncertainties in the data and the sensitivity to film quality appear to be such that these results should be considered consistent with one another. Both groups used the 3ω technique to measure λ.

Some of the first λ measurements for a superlattice were performed on symmetrical GaAs/AlAs multilayers; individual layer thickness was in the range of 5 to 50 nm [8.19]. Thermal conductivity was estimated from thermal diffusivity measured parallel to layers. It was found that the superlattice λ was much larger than that of the corresponding alloy for 50-nm layers, but only ~ 20% larger for 5.0-nm layers (Fig. 8.5). These results contrast with both the Bi$_2$Te$_3$–Sb$_2$Te$_3$ and Si–Ge data, in which λ values lower than those of the equivalent alloys were measured.

Although the focus in nonquantum well superlattice studies of thermoelectric materials is typically on reducing λ, Okamoto et al. [8.33] reported superb electronic properties. An anomalously large (0.2 W m^{-1} K^{-2} at 400 K) power factor was

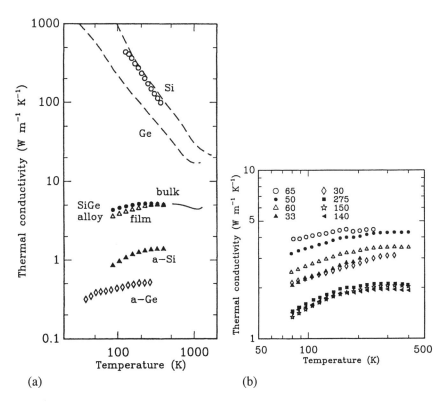

Fig. 8.4. Thermal conductivity of MOCVD Si_mGe_n superlattices of varying periods (**b**). The periods are given in angstroms, and m and n refer to monolayers, ~ 1.38 Å each. The compositions of the superlattices are as follows (period:mxn): 65:22x25, 50:23x13, 60:35x9, 33:19x5, 30:13x9, 275:146x54, 150:74x34, 140:72x30. (**a**) compares the thermal conductivity of SiGe alloys in bulk and thin film forms to confirm the reliability of the 3ω method in this application. Reprinted from [8.18]. Copyright 1997, American Institute of Physics

found for a fifty-two-period, 0.9-nm Si/3.0-nm Ge superlattice. In these structures, the Ge is heavily Au-doped (9 wt%), and the Si is undoped. The reported electrical properties at 400K are $\alpha = 20$ mV K^{-1} and $\rho = 0.2$ Ω cm. We note, though, that the band gap of Si is approximately 1 eV, and so a report of $\alpha \gg 1$ mV K^{-1} at 400 K [8.33] is surprising.

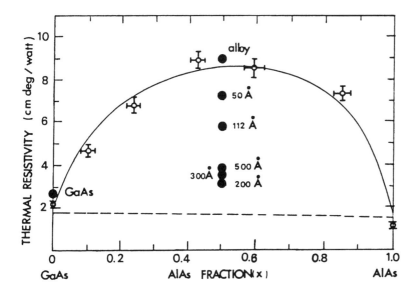

Fig. 8.5. Thermal resistivity of $Al_xGa_{1-x}As$ alloys and AlAs/GaAs superlattices as a function of Al molar fraction. The thickness given is for the individual AlAs and GaAs layers, which were equal. The layers were grown by MBE on (100) GaAs. The results are for heat flow parallel to the layers and were derived from diffusivity measurements (after substrate removal) using the volumetric specific heat of an $Al_{0.5}Ga_{0.5}As$ alloy. Reprinted from [8.19]. Copyright 1987, American Institute of Physics

8.3.3 Other Superlattices

Harman et al. [8.34] report intriguing results for a novel superlattice in the PbTe/Te system. A Te layer only two or three monolayers thick is adsorbed on a PbTe surface. No significant degree of quantum confinement can result from such a thin barrier layer, yet $\alpha^2\sigma$ values 25% greater than bulk PbTe were measured. A potential barrier model was applied to the data, but the conclusion was that a simple change in the effective scattering parameter more closely fits the Seebeck, magnetoresistance, and Hall mobility data [8.35].

Another interesting superlattice variant was reported for the PbTe/PbSe system. The growth mode for $PbSe_{0.98}Te_{0.02}$ on PbTe is that a thin wetting layer forms, followed by the emergence of islands (Stanksi–Krastanov growth mode [8.36]). The thickness of the wetting layer is 0.6 to 0.7 nm. The next layer of PbTe replanarizes the surface. This pattern can be repeated for 600+ periods of $PbTe/PbSe_{0.98}Te_{0.02}$. These 2-D/0-D structures showed the highest α for a given carrier density yet seen in MBE Pb salt superlattice studies. The enhanced electrical properties are attributed to a more favorable carrier scattering mechanism and to partial confinement of electrons [8.37].

8.3.4 Lithographic Quantum Wires

Typically, quantum wires have been made by a suitable combination of epitaxy and lithography resulting in, for instance, GaAs wires embedded in $Ga_{0.65}Al_{0.35}As$ [8.38]. If such techniques can be extended to fabricate successive layers of wires and barriers in materials with low λ, then there might be some applicability to thermoelectric cooling in the same situations where 2-D quantum well superlattices could be useful. At present, however, interest is focused on an entirely different means of synthesizing 1-D thermoelectric materials, filling arrays of channels in thin templates.

8.3.5 Nanowires in Templates

The basic idea is that a high density of parallel submicron diameter channels are created in a substrate and then filled with the thermoelectric material (Fig. 8.6). The lengths of the channels and, consequently, the resulting array of wires are normal to the substrate. Anodic alumina (Fig. 8.7) [8.39, 8.40] and mica [8.41] wafers have been used as templates to support the formation of Bi [8.39–8.41], Sb [8.42], and Bi_2Te_3 [8.42] nanowires. These are suitable subtrates because they are relatively inert chemically and have low λ_L, 1 to 2 W m^{-1} K^{-1} for anodic alumina and 0.8 W m^{-1} K^{-1} for mica [8.41]. Wires can be formed in the template pores by pressure injection from a molten source [8.40], by vacuum evaporation [8.43] (Fig. 8.8), or by electrochemical deposition [8.41]. Extreme aspect ratios can be achieved by pore filling with wire lengths approaching 0.1 mm for 80 nm diameters, and wire diameters as low as 8 nm for 10-micron lengths [8.41]. Typical filling factors have been in the range of 10–20%, but templates with 200-nm diameters and 50% pore fraction are available [8.40].

Measurements of the transport properties of nanowire arrays are in the initial stages. With Bi, an array of 200-nm diameter, 90-micron long wires has a 300 K resistivity of 5 x 10^{-3} Ωcm after taking account of the filling fraction. (The resistivity of bulk Bi is approximately 0.12 x 10^{-3} Ω cm.) Much larger resistivity values were found for aggregates of smaller diameter wires [8.43], but it was concluded that only a minute fraction of the wires contributed to the electrical conduction. Very recently, individual Bi nanowires have been successfully isolated for electrical measurements (Fig. 8.9). The ρ obtained for a 70-nm diameter wire at room temperature was approximately six times greater than the bulk value [8.44]. The resistance of arrays of Bi wires with 200-nm [8.40] and 40-nm diameters [8.44] decreases and increases respectively, as the temperature is

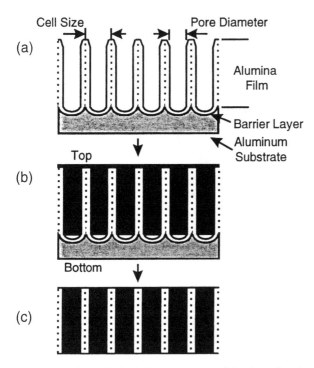

Fig. 8.6. Schematic depiction of pressure injection of molten metal (**b**) into a porous anodic alumina template (**a**) to form a nanowire array (**c**). Reprinted from [8.39] with permission

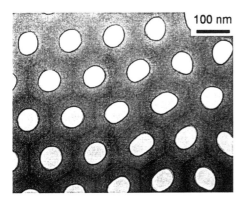

Fig. 8.7. Transmission electron microscope plan-view image of an anodic alumina template with average pore diameter of 56 nm and pore density of 7.4×10^9 cm^{-2}. Reprinted from [8.39] with permission

250   8. Low Dimensional Thermoelectric Materials

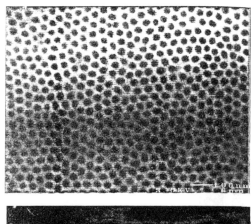
(a)

(b)

Fig. 8.8. The lower panel (**b**) shows a nanowire array formed by vacuum evaporation of Bi onto an anodic alumina substrate with 200-nm pore diameter. The upper panel (**a**) shows an unfilled template with a much smaller pore diameter of 28 nm. Reprinted from [8.43]. Copyright 1998, American Physical Society

lowered, perhaps signaling a semimetal-to-semiconductor transition, as the nominal wire diameter decreases. Similarly, Zhang et al. [8.45] observed changes in magnetoresistance temperature dependence when comparing samples with average wire diameters of 65 and 109 nm.

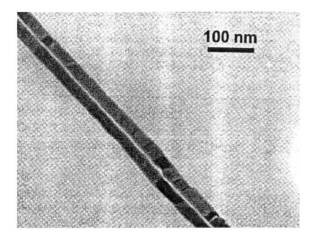

Fig. 8.9. Transmission electron microscope image of 23-nm diameter Bi nanowires. The nanowires were isolated by dissolving the alumina template in an acid solution. Reprinted from [8.39] with permission

8.4 Practical Considerations

8.4.1 Heat leaks

There are various levels at which one can discuss the zT of quantum well superlattices. The first is the zT of the quantum well itself, which neglects the thermal conductance of the barrier layers but is useful for showing the impact of size effects on the active material. The second is the zT of the entire superlattice, which could conceivably be measured in a laboratory. The third is the zT of the superlattice plus substrate, which is probably the best that one could hope to put into practice. Presumably, the substrate plays no role electrically, and zT will be reduced by the ratio of the superlattice thermal conductance to the combined thermal conductance of the superlattice and substrate.

What is required is a mechanically robust material with a low thermal conductivity. One suggestion that has been offered is amorphous SiC [8.46]. This material can be deposited onto a Si wafer, and then the wafer is removed before depositing the superlattice (or simple thin films). The thermal conductivity of amorphous SiC is reportedly 2.0 W m^{-1} K^{-1} [8.46], so a substrate as thick as the superlattice will reduce zT by more than 50% if the superlattice has a thermal conductivity comparable to today's best materials. However, it may be possible to make the SiC membrane significantly thinner than the superlattice/thin film thermoelectric elements [8.47].

For thermoelectric applications, a lithographic quantum wire would be subject to thermal shorting, as just discussed. This problem is mitigated somewhat by the filled template approach, if one chooses a template with low λ_L, such as anodic alumina. It is possible to dissolve such templates [8.39], leaving only air to thermally short the nanowires, but then one is faced with the fragility of the nanowires and their tendency to oxidize.

8.4.2. Heat Pumping

In any event, it is difficult to imagine a single "in-plane" thermoelectric branch with an area-to-length ratio that approaches typical values for bulk thermoelectric elements. Consequently, applications that require minimal heat pumping are the most favorable. A large number of low-current branches in parallel (as with nanowire arrays) or a high density of thermoelectric junctions will be required to achieve significant heat pumping.

In practice, one would expect that a single thermoelectric element would be composed of a patch of nanowires, so that a reasonable current could flow through the device. Electrochemical deposition allows convenient patterning of the process, and alternating p-type and n-type patches could be linked together to form a device.

8.4.3 Heat Spreading

One advantage of the "cross-plane" approach is that it is possible to fabricate thermoelectric elements with a familiar area-to-length ratio but with a much higher density of junctions. The device would have a convenient current requirement but could in principle pump much more heat per unit area than a conventional device. However, it is necessary to have efficient heat spreading and dissipation on the hot side, or the scheme fails.

For instance, a 60 x 60-μm^2 Bi_2Te_3/Sb_2Te_3 element 10 μm long (corresponding, for example, to a superlattice thickness) has nearly the same optimum current as the branches in common cooler designs. Suppose, then, that a device is fabricated with such elements spaced 0.1 mm apart. (Such a device would be limited in size to nominally 0.1 cm^2 to maintain a voltage requirement less than 100 V.) At a current of 2 A and a ΔT of 60°C, the heat pumping capacity will be ~ 300 Wcm^{-2}, and the heat dissipation requirement will be ~1400 W cm^{-2}. A typical aluminum fin design with an area of 25 cm^2 will experience a temperature rise of ~ 1°C W^{-1} above ambient with forced-air cooling. This approach to heat dissipation clearly is not suitable if thermoelectric microcoolers are to advantageously used for high-capacity applications.

8.4.4 Contact Resistance

Perhaps the most serious problem that must be dealt with in short elements is contact resistance. The product of the resistivity and length of the thermoelectric branch, ρl, sets the scale for estimating the significance of contact resistance ρ_c. To fully exploit the zT of the low-dimensional material, we must have $\rho_c \ll \rho l$. For elements fabricated from a layered structure with a total thickness of $l = 10$ μm and $\rho = 10^{-3}$ Ω cm, $\rho_c \ll 10^{-6}$ Ω cm^2. Such low ρ_c values can be achieved in theory but may be difficult to obtain in practice. Similarly, a contact resistance problem may be encountered in constructing devices from nanowire arrays. But with wire lengths in the vicinity of 0.1 mm demonstrated to date, the problem may be manageable.

8.4.5 Bulk Synthesis

The practical problems associated with low-dimensional materials would be greatly reduced if it were possible to fabricate them in bulk quantities. Relative to quantum wells grown by MBE, one of the attractive aspects of the nanowire array approach is that it appears more amenable to manufacturing devices that can pump heat at the levels required for many applications (~1 Wcm^{-2}). Investigations are underway to find other schemes that can produce low-dimensional materials in the quantities and aspect ratios needed.

Intercalation of layered materials is a proven technique for creating 2-D electrical transport in some materials systems [8.48]. Typically small single crystals of the layered material are made to absorb the intercalate from solution, leading to a large expansion of the crystals in the direction normal to the layers of the host structure. When the intercalate is such that it will form an insulating layer, there is the possibility of creating macroscopic quantum well structures. If the intercalate is a large band-gap material whose layer thickness is greater than about 1.0 nm, one can expect to see a transformation of electrical properties that reflects a partial or complete switch to 2-D conduction. There have not been any reports of successful application of this concept to thermoelectric materials, but work is underway that could lead to such a development [8.49].

Nature offers another example of layered bulk materials in the structure of certain two-phase solids directionally crystallized from eutectic compositions [8.50]. The repeat distances in such phases are, of course, not as precise as one would obtain synthetically, but it is possible to obtain a narrow distribution of layer thickness and to control the average layer thickness by controlling the crystallization conditions. This approach seems more suited to creating structures with 1-μm layers, rather than 1-nm layers.

8.5 Summary

Interest in low-dimensional thermoelectric materials has arisen because theory and experiment indicate performance advantages compared to bulk materials. The beneficial effect of confinement with respect to electrical properties has been demonstrated experimentally. Band-edge engineering has been applied theoretically to practical materials systems, and the results encourage experimental efforts. There are also numerous theoretical and experimental indications that λ_L is decreased by scattering or reflection from boundaries or by changes in the average group velocity of the phonons.

To capture the performance advantage in a laboratory device, the design must minimize the impact of both parasitic heat flow and contact resistance. Further issues then arise as to whether the device can be manufactured cost-effectively. The search continues for low-dimensional materials for which the physics, engineering, and economics could mesh into a practical device.

9. Thermionic Refrigeration

The flow of current that gives rise to thermoelectric effects is diffusive. However, it is possible to obtain similar effects when ballistic flow takes place. This distinction is easily recognized when we compare the current in a vacuum diode with that in a bulk semiconductor. Ballistic type of flow is then called thermionic transport. This type of transfer can also take place in the solid state, and it is possible for diffusive effects in a semiconductor to give way to thermionic effects when the effective length between two barriers becomes very small [9.1]. Although there does not yet seem to be a practical refrigerator based on thermionic effects, there are good reasons for thinking that such a device will appear in the not too distant future. Thus, in this chapter, we discuss the basic physics of thermionic refrigeration. We treat transport both in a vacuum and in the solid state.

9.1 The Vacuum Diode

Ioffe [9.2] recognized the advantage of an energy conversion device in which heat transfer takes place across a vacuum. He pointed out that, in principle, the only unwanted heat transfer between the anode and cathode in a thermionic diode would be by radiation. For Ioffe, this would not have been a negligible effect because he had in mind thermionic generators with a heat source at 1000 K or more. He dismissed the idea of energy conversion using thermionic diodes at lower temperatures because he knew of no cathode materials with a work function smaller than about 1 eV, and an emission current below about 1000 K would have been too small for practical purposes.

Mahan [9.3] was not quite so pessimistic and thought that a thermionic refrigerator operating at a temperature of around 500 K might be possible. This would require a work function of not more than 0.7 eV. Although the applications of the refrigerator envisaged by Mahan would be rather limited, he deemed it worthwhile to develop the theory of the device. This theory is, of course, also applicable at lower temperatures.

Now, it seems possible that practical materials with work functions as low as 0.3 eV may eventually become available. For example, very low work functions have been reported for alkalides and electrides [9.4]. Measurable electron emission was observed at a temperature as low as 193 K [9.5], and a work function of no more than 0.2 eV was mentioned. As we shall see, thermionic refrigeration at ordinary temperatures will present certain practical problems, even if satisfactory cathode materials are developed. Nevertheless, the technique offers real prospects for a substantial improvement in the coefficient of performance.

Figure 9.1 shows a schematic plot of potential energy against displacement for a thermionic diode. Mahan considered the case of different work functions for the anode and cathode, but we shall suppose that the two work functions are equal to Φ. A voltage V is applied between the two electrodes so as to make the electron current from the cathode exceed that from the anode. The horizontal lines represent the chemical potential within the electrodes.

The basic equation for thermionic emission is due to Richardson [9.6] and may be written as

$$J_n = A_0 T_n^2 \exp\left(-\frac{\Phi}{k_B T_n}\right), \tag{9.1}$$

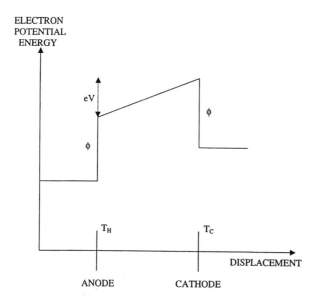

Fig. 9.1. Schematic plot of electron potential energy against displacement for a thermionic diode carrying an electric current from the cold cathode to the hot anode. Reprinted from [9.6]. Copyright 1999, American Institute of Physics

where J_n is the emitted current density and the subscript n is used to denote a particular electrode. For the constant A_0, we assume the ideal value of $4\pi e m k_B^2/h^3$, which is equal to 1.2×10^6 A m^{-2} K^{-2}.

Part of the applied voltage, equal to V_0, is needed to balance the effect that the anode has a higher temperature than that of the cathode. The remaining part, $V - V_0$, actually drives the current. Taking as positive the direction of the conventional current into the cathode, the total current that flows in the space between the electrodes is

$$J = J_C - J_H = A_0 \left\{ T_C^2 \exp\left(-\frac{\Phi}{k_B T_C}\right) - T_H^2 \exp\left(-\frac{\Phi + eV}{k_B T_H}\right) \right\}. \tag{9.2}$$

Note that V_0 is the voltage that must be applied for the current to be equal to zero and, in this sense, it has some resemblance to the Seebeck voltage in a conventional thermocouple. Its value can be found from Eq. (9.2), where $J = 0$. It is given by

$$eV_0 = \Phi\left(\frac{T_H}{T_C} - 1\right) + 2\ln\left(\frac{T_H}{T_C}\right). \tag{9.3}$$

Now, we may determine the cooling effect of the current. The electrons that leave the cathode require a certain amount of energy to do so. This energy consists partly of a potential term equal to Φ and partly of a kinetic term equal to $2k_B T$. Thus, when the electrons leave the cathode, this electrode loses heat at the rate q_J per unit area, where q_J is given by

$$q_J = J\frac{(\Phi + 2k_B T)}{e}. \tag{9.4}$$

The rate of energy expenditure per unit area is given by

$$w_J = JV. \tag{9.5}$$

Because the space between the electrodes consists of a vacuum, there is no heat transfer by conduction or convection, but thermal radiation will take place. The rate of radiation per unit area is given by

$$q_R = -\varepsilon\sigma\left(T_H^4 - T_C^4\right), \tag{9.6}$$

where ε is the thermal emissivity of the electrode surfaces and σ is the Stefan–Boltzmann constant. The negative sign indicates that thermal radiation opposes cooling of the cathode.

The vacuum diode that has been described acts in more or less the same way as a negative thermoelement. It is, of course, necessary to provide a return path for the electric current, and this implies a reduction in the coefficient of performance due to electrical resistance and thermal conductance [9.6]. Although, in principle, the return path could take the form of a positive thermoelement, it would probably be best to use a passive metallic conductor in which the ratio of thermal to electrical conductivity would be given by the Wiedemann–Franz law. If R and K, respectively, are the electrical resistance and thermal conductance of the passive conductor, corresponding to the unit cross-sectional area of the diode, the overall cooling effect is

$$q = q_J + q_R - J^2 R/2 - K(T_H - T_C), \qquad (9.7)$$

and the rate of expenditure of electrical energy is

$$w = w_J + J^2 R. \qquad (9.8)$$

Before we derive the coefficient of performance (COP) of the thermionic diode, we must discuss an effect that will certainly present practical difficulties, the effect of space charge in the vacuum region. At any time, there will be a substantial concentration of free electrons between the electrodes, and these will tend to set up a barrier that opposes emission from the cathode. This problem has been discussed in some detail by Mahan [9.3], and he concluded that, for a device working at 700 K, the space charge effect can be overcome only by reducing the distance between the anode and cathode to about 1 mm or less. Because the space charge problem becomes more acute as the temperature is reduced, it is clear that we must take it into account.

Mahan has shown that the potential function in the space between the electrodes can be solved analytically when symmetry exists, that is, the temperatures, chemical potentials, and work functions are the same at both electrodes. Although we expect the temperatures to be different, the analytical solution suffices for our purposes. The potential function is

$$U(x) = -2k_B T \ln[\cos(ax)], \qquad (9.9)$$

where

$$a = p \exp\left[-(eV_m - \zeta)/2k_B T\right], \qquad (9.10)$$

where V_m is the maximum potential and

$$p^2 = \frac{4\pi e^2}{k_B T \lambda^3}, \qquad (9.11)$$

where

$$\lambda = \left(\frac{2\pi\hbar^2}{mk_B T}\right)^{1/2}. \qquad (9.12)$$

To avoid a divergence in Eq. (9.9), it is necessary that $ad < \pi$, where d is the spacing between the electrodes. Then, this sets the highest potential at a value given by

$$V_m - \frac{\zeta}{e} > \frac{2k_B T}{e} \ln\left(\frac{pd}{\pi}\right). \qquad (9.13)$$

Here p is defined in terms of a number of universal constants and the absolute temperature T. At 300 K, Mahan showed that an electrode spacing d, equal to 1 mm, yields a space charge barrier of at least 0.71 eV. This means that p is equal to 2.9 x 10^5 m^{-1}. If the space charge barrier is not to exceed 0.3 eV, the electrode spacing has to be reduced to no more than about 0.4 µm. It is clear that operation near room temperature will require the anode and cathode to be very close together and an appropriate technique will be needed to maintain the necessary separation. Let us suppose, however, that such a technique can be developed.

The coefficient of performance ϕ is equal to q/w and can be obtained from Eq. (9.7) and (9.8). It is obvious that, whatever the theoretical coefficient of performance, a thermionic refrigerator will be useful only if it has reasonable cooling power. It is suggested that the smallest acceptable value for q is about 10^4 W m^{-2}. In Fig. 9.2, we plot the largest value of the cooling power against the work function, and neglect losses in the passive conductor. The cathode is supposed to be at 260 K and, when radiation losses are included, the anode is at 300 K; the emissivity is taken to be equal to unity. Note that the radiation losses are almost negligible when Φ is less than about 0.35 eV. Our arbitrary lower limit for the cooling power also requires that Φ be not much greater than 0.3 eV. If we take account of the losses in the passive conductor, the upper limit for the

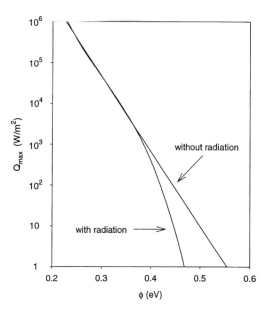

Fig. 9.2. The largest cooling power Q_{max} for a thermionic diode in the absence of losses in the passive conductor. The cathode is at 260 K and, when radiation losses are included, the anode is taken to be at 300 K. Reprinted from [9.6]. Copyright 1999, AIP

work function of the cathode is indeed about 0.3 eV, and we shall use this value in subsequent calculations.

Now, we consider optimizing the passive conductor. We can do this most easily for the condition of maximum cooling power. This requires that the part of the Joule heating, $J^2R/2$, that reaches the heat source is equal to the heat conduction, $K(T_H - T_C)$. Using the Wiedemann–Franz relationship, this gives

$$R = \frac{\{L(T_H - T_C)(T_H + T_C)\}^{1/2}}{J}, \qquad (9.14)$$

where L is the Lorenz number. The losses associated with the passive conductor do not have a great effect on the cooling power, but they are quite significant in calculating ϕ because the thermionic diode is inherently such an efficient energy converter that the energy expenditure in the device itself is far less than the heat that is pumped. Thus, to obtain the maximum coefficient of performance, the electrical resistance should be substantially larger than the value given by Eq. (9.14). An analytical expression for the optimum resistance is difficult to obtain because of the exponential terms in the relationship between current and applied voltage.

Figure 9.3 shows how the maximum temperature difference varies with the work function of the cathode, if the heat sink is maintained at 300 K. Note that the temperature can be lowered by about 100 K when the value of Φ is 0.3 eV. However, the real virtue of the thermionic refrigerator would be in obtaining very high values for ϕ at smaller temperature differences.

In Fig. 9.4, it is shown how cooling power and ϕ vary with applied voltage in a specific case. Again it is assumed that $\Phi = 0.3$ eV and that the source and sink temperatures are 260 K and 300 K, respectively. For both the conditions illustrated, the cooling power is well in excess of 10^4 Wm^{-2} and ϕ is greater than three over the range of voltage that might be employed.

The fact that there is a maximum in the plot of the coefficient of performance against voltage is due to radiation losses, small though they are. The cut-off in both cooling power and ϕ at just over 0.05 V is also due to the fact that a lower applied voltage is insufficient to compensate for the higher temperature of the heat sink, compared with that of the source.

It is interesting to compare the performance of a hypothetical thermionic refrigerator with that of a thermoelectric refrigerator. For this comparison, we assume that a thermoelectric figure of merit ZT equal to four might eventually be achieved. Figure 9.5 shows that the thermionic refrigerator would have the superior ϕ for heat source temperatures in excess of about 240 K.

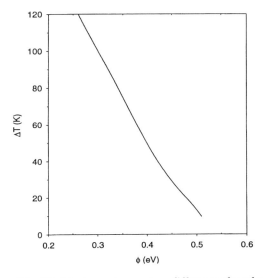

Fig. 9.3 Maximum temperature difference plotted against work function for a thermionic refrigerator operating at $T_H = 300$ K. Reprinted from [9.6]. Copyright 1999, AIP

262   9. Thermionic Refrigeration

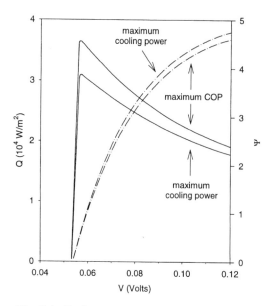

Fig. 9.4 Cooling power (solid line) and coefficient of performance (COP or Ψ, dash-dot line) of a thermionic refrigerator plotted against applied voltage for $\Phi = 0.3$ eV, $T_H = 300$ K, and $T_C = 260$ K. Reprinted from [9.6]. Copyright 1999, AIP

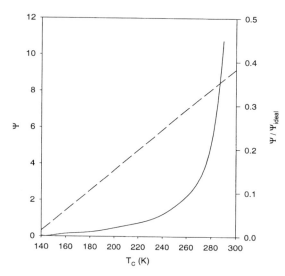

Fig. 9.5 Ratio of the coefficient of performance Ψ to that of a Carnot cycle for a thermionic refrigerator (solid line) with $\Phi = 0.3$ eV and a thermoelectric refrigerator (dashed line) with $ZT = 4$. Reprinted from [9.6]. Copyright 1999, AIP

9.2 Solid-State Thermionic Devices

Now, we consider the possibility of thermionic converters in which the emitted charge carriers traverse a solid medium rather than a vacuum. This means that we lose one of the principal attractions of the vacuum diode, namely, the absence of heat conduction losses. However, by way of compensation, we overcome some of the problems associated with the vacuum device. It should not be difficult, in principle, to establish the barrier height at any desired value. Again, it should be possible to bring the electrodes close enough to avoid space charge problems. Indeed, as we shall see, close spacing is essential if thermionic rather than thermoelectric operation is desired. A further advantage of the solid-state converter is that transport by both electrons and holes is possible, so that one should be able to eliminate the passive conductor that is a necessary part of the vacuum device.

Thermionic refrigeration has been achieved by Shakouri et al. [9.7] using a single-barrier heterostructure. These authors constructed a device consisting of a 1-μm In–Ga–AsP barrier between a 0.3-μm cathode and a 0.5-μm anode, both made from n⁺-In–GaAs. The cooling of the cathode was only 0.5° at a temperature of 20°C. Nevertheless, it was claimed that cooling through 10 K should be possible with an improved design, and it was stated that theoretical predictions forecast a temperature difference of up to 40 K. Shakouri et al. [9.7] mentioned that they had performed experiments to establish that the cooling was due to thermionic emission and not to the Peltier effect.

Mahan et al. [9.1] gave a theoretical treatment of thermionic energy conversion. They showed that the temperature difference across a single barrier should be small; otherwise, there would be unacceptable heat conduction losses. However, a multilayer configuration should allow maintaining large temperature differences between the source and sink. First, we shall discuss a single barrier and then consider the multilayer arrangement. As in the treatment by Mahan and his colleagues, we shall assume that the carriers have positive signs.

For a single barrier, the theory is similar to that for a vacuum diode. We must, however, replace the term for thermal radiation loss by one that takes account of heat conduction. Also, we need not include the losses in the passive conductor. It is assumed that a negligible number of the charge carriers undergo collisions in the barrier region, but the latter must be large enough to prevent tunneling effects. The maximum thickness d_t for tunneling to occur may be expressed as

$$d_t = \frac{\hbar}{2k_B T}\sqrt{\frac{e\Phi}{m^*}}. \tag{9.15}$$

Then, we require a barrier thickness d that satisfies the inequality

$$l_e > d > d_t, \tag{9.16}$$

where l_e is the mean free path of the carriers. Because d_t is usually less than 10 nm and l_e can be an order of magnitude greater, it is probably not difficult to meet this condition.

If $1/K$ is the thermal resistance of the barrier, including any interfacial contributions, the cooling effect is

$$q = q_J - K\Delta T, \tag{9.17}$$

where q_J is given by Eq. (9.4) and ΔT is the temperature difference across the barrier. Now, because the barrier must be very thin, K is likely to be large, and the thermal conduction losses will outweigh the thermionic cooling, unless ΔT is small. For this reason, Mahan and his co-workers proposed a multilayer device. However, for the time being, we shall continue to assume that there is only one barrier.

The fact that ΔT is small compared with the temperatures T_H and T_C simplifies the theory to some extent. Thus, the subscript n may be dropped in Eq. (9.1), so that, at temperature T,

$$J_T = A_0 T^2 \exp\left(-\frac{\Phi}{k_B T}\right), \tag{9.18}$$

and Eq. (9.2) then becomes

$$J = \frac{eA_0 T}{k_B}(V - V_0)\exp\left(-\frac{\Phi}{k_B T}\right), \tag{9.19}$$

where V is the applied voltage over the single barrier. V_0 is given by

$$V_0 = \frac{k_B \Delta T}{e}(b+2), \tag{9.20}$$

where

$$b = \frac{\Phi}{k_B T}. \tag{9.21}$$

Note that, in the solid-state, it is possible for Φ to be small enough so that classical statistics becomes invalid. However, it turns out that, if the coefficient of performance is optimized and if its value is to exceed that which can be achieved with established thermoelectric converters, then classical statistics may be used.

The cooling power per unit area is

$$q = J_T(b+2)(V - V_q), \qquad (9.22)$$

where

$$V_q = \frac{k_B \Delta T}{e}(b+2+c). \qquad (9.23)$$

The quantity c is defined as

$$c = \frac{2 + eK/k_B J_T}{b+2}. \qquad (9.24)$$

Thus, ϕ is

$$\phi = \frac{q}{w} = \frac{k_B T(b+2)}{e} \frac{(V - V_q)}{V(V - V_0)}. \qquad (9.25)$$

Mahan and his colleagues showed that ϕ reaches its maximum value ϕ_{max}, when the voltage is given by

$$V_\phi = V_q + \sqrt{V_q(V_q - V_0)}, \qquad (9.26)$$

and they found that

$$\phi_{max} = \frac{T}{\Delta T} \frac{b+2}{\left(\sqrt{b+2+c} + \sqrt{c}\right)^2}. \qquad (9.27)$$

The first term on the right-hand side of this equation is the coefficient of performance of a Carnot cycle, and the second term represents the factor by which ϕ is degraded.

266   9. Thermionic Refrigeration

A parameter T_R that has the dimensions of temperature is defined by the relationship

$$\left(k_B T_R\right)^2 = \frac{2\pi^2 \hbar^3 K}{m^* k_B T}. \qquad (9.28)$$

The coefficient of performance for given temperatures of the source and sink is then a function of T_R and Φ. In Fig. 9.6, the ratio of the coefficient of performance to that of a Carnot cycle is plotted against the work function for various values of T_R. It is clear that T_R should be as small as possible and that, if the coefficient of performance is to reach a reasonably large fraction of the ideal value, the barrier height Φ should be somewhat larger than $k_B T$.

If the temperature of the heat source is to be lowered by a worthwhile amount, it is necessary to use a multilayer arrangement. Then, it is necessary to calculate the overall performance using the same principles that are employed for a thermoelectric cascade. Mahan and his co-workers carried out this calculation, and their results are shown in Fig. 9.7 for a heat source at 260 K and sink at 300 K. They noted that the coefficient of performance is about the same as that for a thermoelectric refrigerator whose ZT is equal to unity, when $T_R = 500$ K. Thus,

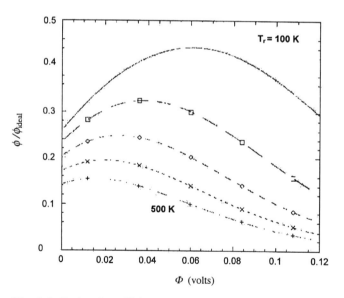

Fig. 9.6 Ratio of coefficient performance to that of an ideal refrigerator for a single barrier, as a function of the barrier height Φ for values of T_R between 100 and 500 K. Reprinted from [9.1]. Copyright 1999, American Institute of Physics

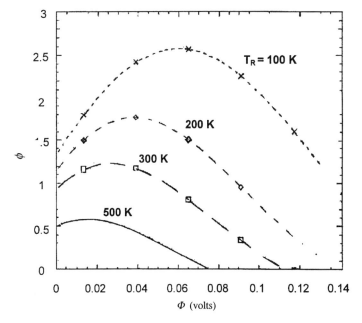

Fig. 9.7 Coefficient of performance of a multilayer thermionic refrigerator, where $T_C = 260$ K and $T_H = 300$ K. The variables are the barrier height Φ and the parameter T_R. Reprinted from [9.1]. Copyright 1999, American Institute of Physics

significantly smaller values of T_R are needed if the multilayer thermionic device is to be advantageous.

Mahan et al. [9.1] drew attention to an interesting aspect of solid-state thermionic refrigeration. They showed that, when the voltage and temperature difference across each barrier is small, the equations become identical with those of a thermoelectric refrigerator if appropriate substitutions are made. These substitutions are

$$\alpha_{TI} = \frac{k_B}{e}(b+2), \tag{9.29}$$

$$\sigma_{TI} = \frac{eJ_T d}{k_B T}, \tag{9.30}$$

$$\lambda_{TI} = \left(2\frac{k_B}{e}J_T + K\right)d, \tag{9.31}$$

and

$$z_{TI}T = \frac{\alpha^2 \sigma T}{\lambda} = \frac{b+2}{c}. \tag{9.32}$$

This suggests that, although solid-state thermionic converters are distinctly different from thermoelectric refrigerators and generators, there may be a close connection between them. That this is indeed the case has been shown by Vining and Mahan [9.8].

Chasmar and Stratton [9.9] defined a parameter β to describe the suitability of a given semiconductor for thermoelectric applications. Vining and Mahan [9.8] pointed out that a similar parameter β_{TI} can be defined in the thermionic case as

$$\beta_{TI} = \frac{m^* k_B (k_B T)^2 d}{2\pi^2 \hbar^3 \lambda_L}. \tag{9.33}$$

Then, the equation

$$z_{TI}T = \frac{(b+2)^2}{2 + (\exp b)/\beta_{TI}} \tag{9.34}$$

is similar to the equation that gives zT for a nondegenerate thermoelectric material. This means that the thermionic converter will not have a higher coefficient of performance or efficiency than a thermoelectric converter unless $\beta_{TI} > \beta$.

Vining and Mahan [9.8] showed that the ratio of β_{TI} to β is given by

$$\frac{\beta_{TI}}{\beta} = \frac{d}{l_e \sqrt{\pi}}, \tag{9.35}$$

where d is the barrier thickness and l_e is the mean free path of the charge carriers. Now, we have already assumed that $d < l_e$, so that any given material will give better performance when used in the thermoelectric mode rather than the thermionic mode. However, this assumes that the lattice thermal conductivity λ_L is the same in both cases, but it has been shown that the λ_L falls well below its bulk value in thin layers of material [9.10]. Therefore, it is possible for β_{TI} to exceed β in spite of the fact that d is necessarily smaller than $l_e \pi^{1/2}$.

In conclusion, it appears that thermoelectric effects are inherently superior to thermionic effects, but it may be necessary to operate in the thermionic regime to gain the benefit of a reduction in lattice conductivity that comes with small thickness.

References

Chapter 1

1.1 G. Borelius, W.H. Keesom, C.H. Johansson, J.O. Linde: Proc. Ned. Akad. Wet **35**, 10 (1932)
1.2 J.W. Christian, J.P. Jan, W.B. Pearson, I.M. Templeton: Can. J. Phys. **36**, 627 (1958)
1.3 B.J. O'Brien, C.S. Wallace: J. Appl. Phys. **29**, 1010 (1958)
1.4 R.T. Delves: Br. J. Appl. Phys. **15**, 105 (1964)
1.5 R.B. Horst: J. Appl. Phys. **34**, 3246 (1963)

Chapter 2

2.1 J.F. Nye: *Physical Properties of Crystals* (Clarendon, Oxford Press 1957)
2.2 R. Berman, F.E. Simon, J.M. Ziman: Proc. R. Soc., London, Ser. A **220**, 171 (1953)
2.3 P. Debye: Ann. Phys. **39**, 789 (1912)
2.4 B.N. Brockhouse: Phys. Rev Lett. **2**, 256 (1959)
2.5 A. Eucken: Ann. Phys. **34**, 185 (1911)
2.6 P. Debye: *Vorträge über die kinetische Theorie* (Teubner), (1914)
2.7 R.E. Peierls: Ann. Phys. **3**, 1055 (1929)
2.8 H.B.G. Casimir: Physica **5**, 495 (1938)
2.9 H.J. Goldsmid, A.W. Penn: Phys. Lett. A **27**, 523 (1968)
2.10 T.H. Geballe, G.W. Hull: Phys. Rev. **110**, 773 (1958)
2.11 J. Callaway: Phys. Rev. **113**, 1046 (1959)
2.12 J.E. Parrott: Proc. Phys. Soc. **81**, 726 (1963)
2.13 A. Sommerfeld: Z. Phys. **47**, 1, 43 (1928)
2.14 A.F. Ioffe: *Physics of Semiconductors* (Infosearch, London 1960)
2.15 B.R. Nag: *Electron Transport in Compound Semiconductors* (Springer, Berlin, Heidelberg 1980)
2.16 A.C. Beer: *Galvanomagnetic Effects in Semiconductors* (Academic Press, New York 1963)
2.17 J. McDougall, E.C. Stoner: Philos. Trans. A **237**, 67 (1938)
2.18 P. Rhodes: Proc. R. Soc. A **204**, 396 (1950)

2.19 A.C. Beer, M.N. Chase, P.F. Choquard: Helv. Phys. Acta **28,** 529 (1955)
2.20 P.J. Price: Philos. Mag, **46,** 1252 (1955)
2.21 E.H. Putley: *The Hall Effect and Related Phenomena* (Butterworths Sevenoaks, Kent 1960)
2.22 I.M. Tsidil'kovskii: *Thermomagnetic Effects in Semiconductors* (Infosearch, London 1962)
2.23 R.B. Horst: J. Appl. Phys. **34,** 3246 (1963)
2.24 R.T. Delves: Br. J. Appl. Phys. **15,** 105 (1964)
2.25 J. Kolodzieczak, S. Zhukotynski: Phys. Status Solidi **5,** 145 (1964)
2.26 E.O. Kane: J. Phys. Chem. Solids **1,** 249 (1957)
2.27 T.C. Harman, J.M. Honig: *Thermoelectric and Thermomagnetic Effects and Applications* (McGraw-Hill, New York 1967)
2.28 H.P.R. Frederikse: Phys. Rev. **92,** 248 (1953)
2.29 T.H. Geballe, G.W. Hull: Phys. Rev. **94,** 1134 (1954)
2.30 C. Herring: Phys. Rev. **96,** 1163 (1954)
2.31 L. Gurevich: J. Phys. (USSR) **9,** 477 (1945)
2.32 L. Gurevich: J. Phys. (USSR) **10,** 67 (1946)
2.33 H.J. Goldsmid, C.C. Jenns, D.A. Wright: Proc. Phys. Soc. **73,** 393 (1959)

Chapter 3

3.1 R.P. Chasmar, R. Stratton: J. Electron. Control. **7,** 52 (1959)
3.2 R. Simon: Adv. Energy Convers. **1,** 81 (1961).
3.3 A.F. Ioffe: *Physics of Semiconductors* (Infosearch, London 1960)
3.4 J.D. Wasscher, W. Albers, C. Haas: Solid-State Electron. **6,** 261 (1963)
3.5 R. Simon: Adv. Energy Convers. **3,** 515 (1963)
3.6 K.F. Hulme: In *Materials Used in Semiconductor Devices*, Ed. by C.A. Hogarth (Interscience, New York 1965) p.115
3.7 R.T. Delves: Br. J. Appl. Phys. **13,** 440 (1962)
3.8 J. Bardeen, W. Shockley: Phys. Rev. **80,** 72 (1950)
3.9 E.Z. Gershtein , T.S. Stavitskaya , L.S. Stil'bans: Sov. Phys. Tech. Phys. **2,** 2302 (1957)
3.10 W.A. Harrison: Phys. Rev. **104,** 1281 (1956)
3.11 F.E. Low, D. Pines: Phys. Rev. **98,** 414 (1955)
3.12 D.J. Howarth, E.H. Sondheimer: Proc. Phys. Soc., London, Sect. A **219,** 53 (1953)
3.13 H. Ehrenreich: J. Appl. Phys. **32,** 2155 (1961)
3.14 E.M. Conwell, V.F. Weisskopf: Phys. Rev. **77,** 388 (1950)
3.15 H. Brooks: Adv. Electron. Electron Phys. **7,** 156 (1955)
3.16 A.F. Ioffe: *Semiconductor Thermoelements and Thermoelectric Cooling* (Infosearch, London, 1957)
3.17 R.W. Ure: J. Appl. Phys. **30,** 1922 (1959)

3.18 C. Erginsoy: Phys. Rev. **79,** 1013 (1950)
3.19 W.T. Read: Philos. Mag. **46,** 111 (1955)
3.20 C. Herring: Bell Syst. Tech. J. **34,** 237 (1955)
3.21 C.M. Bhandari, D.M. Rowe: Proceedings, Fifth International Conference on Thermoelectric Energy Conversion (Arlington, Texas 1984), p. 38
3.22 D.M. Rowe, C.M. Bhandari: Proceedings, Fifth International Conference on Thermoelectric Energy Conversion (Arlington, Texas 1984), p. 62
3.23 T.C. Harman, J.M. Honig: *Thermoelectric and Thermomagnetic Effects and Applications* (McGraw-Hill, New York 1967)
3.24 D. Armitage, H.J. Goldsmid: J. Phys. C **2,** 2389 (1969)
3.25 G. Leibfried, E. Schlömann: Nachr. Akad. Wiss., Göttingen, Kl. 2 Math – Physik **2,** 71 (1954)
3.26 J.S. Dugdale, D.K.C. MacDonald: Phys. Rev. **98,** 1751 (1955)
3.27 R.W. Keyes: Phys. Rev. **115,** 564 (1959)
3.28 A.W. Lawson: J. Phys. Chem. Solids **3,** 154 (1957)
3.29 H.J. Goldsmid: *Thermoelectric Refrigeration* (New York, Plenum, 1964)
3.30 H.J. Goldsmid: R.W. Douglas: Br. J. Appl. Phys. **5,** 386, 458 (1954)
3.31 A.V. Ioffe, A.F. Ioffe: Dokl. Akad. Nauk. SSSR **97,** 821 (1954)
3.32 A.F. Ioffe, S.V. Airapetyants, A.V. Ioffe, N.V. Kolomoets, L.S. Stil'bans: Dokl. Akad. Nauk. SSSR **106,** 981 (1956)
3.33 S.V. Airapetyants, B.A. Efimova, T.S. Stavitskaya, L.S. Stil'bans, L.M. Sysoeva: Zh. Tekh. Fiz. **27,** 2167 (1957)
3.34 F.D. Rosi, Solid-State Electron **11,** 833 (1968)
3.35 P.G. Klemens, Phys. Rev. **119,** 507 (1960)
3.36 J.R. Drabble, H.J. Goldsmid: *Thermal Conduction in Semiconductors* (Pergamon, Oxford 1961)
3.37 H.B.G. Casimir: Physica **5,** 495 (1938)
3.38 H.J. Goldsmid, A.W. Penn: Phys. Lett. A **27,** 523 (1968)
3.39 N. Savvides, H.J. Goldsmid: J. Phys. C **6,** 1701 (1973)
3.40 N. Savvides, H.J. Goldsmid: J. Phys. C **13,** 4657, 4671 (1980)
3.41 G.A. Slack, M.A. Hussain: J. Appl. Phys. **70,** 2694 (1991)
3.42 Y.A. Boikov, B.M. Gol'tsman, V.A. Kutasov: Sov. Phys. Solid State **20,** 757 (1978)
3.43 S.L Korolyuk, I.M. Pilat, A.G. Samoilovich, V.N. Slipchenko, A.A. Snarskii, E.F. Tsar'kov: Sov. Phys. Semicond. **7,** 502 (1973)
3.44 V.P Babin, T.S. Gudkin, Z.M Dashevskii, L.D. Dudkin, E.K. Iordanishvili, V.I. Kaidanov, N.V. Kolomoets, O.M. Narva, L.S. Stil'bans: Sov. Phys. Semicond. **8,** 478 (1974)
3.45 A.G. Samoilovich, V.N. Slipchenko: Sov. Phys. Semicond. **9,** 1249 (1975)
3.46 T.S. Gudkin, E.K. Iordanishvili, E.E. Fiskind: Sov. Tech. Phys. Lett. **4** 844 (1978)
3.47 D.K.C. MacDonald, E. Mooser, W.B. Pearson, I.M. Templeton, S.B. Woods: Philos. Mag. **4,** 589 (1959)

3.48 D.K.C. MacDonald, I.M. Templeton: Proceedings, International Conference on Semiconductor Physics, Prague (Academic Press, New York 1960) p. 650
3.49 F.J. Blatt: In *High Magnetic Fields*, Ed. by H. Kolm, B. Lax, F. Bitter, R. Mills (Cambridge, MA: MIT Press; New York, J. Wiley 1961) p. 518
3.50 F.J. Blatt: Philos. Mag. **7,** 715 (1962)
3.51 R.W. Keyes: In *Thermoelectricity: Science and Engineering*, Ed. by R.R. Heikes, R.W. Ure (Interscience, New York 1961) p. 389
3.52 H.J. Goldsmid, A.S. Gray: Cryogenics **19,** 289 (1979)

Chapter 4

4.1 D.J. Ryden: AERE Report 6996, UKAEA, Harwell (1973)
4.2 E.H. Putley: *The Hall Effect and Related Phenomena* (Butterworths, Sevenoaks, Kent, 1960)
4.3 T.C. Harman: J. Appl. Phys. **29,** 1373 (1958)
4.4 H.J. Goldsmid: Proc. Phys. Soc. London. **71,** 633 (1958)
4.5 T.M. Dauphinee, S.B. Woods: Rev. Sci. Instrum. **26,** 693 (1955)
4.6 E. Müller, W. Heiliger, P. Reinshaus, H. Sübmann: Proceedings, Fifteenth International Conference on Thermoelectrics, 412 (1996)
4.7 H.J. Goldsmid: Proc. Phys. Soc. London Sect. B **69.** 203 (1956)
4.8 R. Bowers, R.W. Ure, J.E. Bauerle, A.J. Cornish: J. Appl. Phys. **30,** 930 (1959)
4.9 A.V. Ioffe, A.F. Ioffe: Zh. Tekh. Fiz. **22,** 2005 (1952)
4.10 D.G. Cahill, H.E. Fischer, T. Klitsner, E.T. Swartz, R.O. Pohl: J. Vac. Sci. Technol. **A7,** 1259 (1989)
4.11 D.G. Cahill: Rev. Sci. Instrum. **61,** 802 (1990)
4.12 A.W. Penn: J. Sci. Instrum. **41,** 626 (1964)
4.13 I.S. Lisker: Sov. Phys. Solid State **8** 1042 (1966)
4.14 T.A.A. Al-Obaidi, H.J. Goldsmid: Energy Convers. **9,** 131 (1969)
4.15 W.B. Berry: Energy Convers. **9,** 133 (1969)
4.16 A.E. Bowley, R. Delves, H.J. Goldsmid: Proc. Phys. Soc. London **72,** 401 (1958)
4.17 J.R. Drabble, R. Wolfe: J. Electron. Control **3,** 259 (1957)
4.18 E.H. Volckmann, H.J. Goldsmid: Proceedings, Sixteenth International Conference on Thermoelectrics, Dresden (IEEE, New York 1997) p. 196
4.19 G.L. Guthrie, R.L. Palmer: J. Appl. Phys. **37,** 90 (1966)

Chapter 5

5.1 P.W. Lange: Naturwissenschaften **27,** 133 (1939)
5.2 J.R. Drabble, C.H.L. Goodman: J. Phys. Chem. Solids **5,** 142 (1958)

5.3 G. Offergeld, J. van Cakenberghe: J. Phys. Chem. Solids **11**, 310 (1959)
5.4 H.J. Goldsmid: Proceedings, International Conference on Semiconductor Physics (Academic Press, New York 1960) p. 1015
5.5 J.O. Jenkins, J.A. Rayne, R.W. Ure: Phys. Rev. B **5**, 3171 (1972)
5.6 G.F. Bolling: J. Chem. Phys. **33**, 305 (1960)
5.7 E.C. Itskevitch: Sov. Phys. JETP **11**, 255 (1960)
5.8 R.O. Carlson: J. Phys. Chem. Solids **13**, 65 (1960)
5.9 R.T. Delves, A.E. Bowley, D.W. Hazelden, H.J. Goldsmid: Proc. Phys. Soc. London **78**, 838 (1961)
5.10 J.R. Drabble, R. Wolfe: Proc. Phys. Soc. London Sect. B **69**, 1101 (1956)
5.11 J.R. Drabble: Proc. Phys. Soc. London **72**, 380 (1958)
5.12 J.R. Drabble, R.D. Groves, R. Wolfe: Proc. Phys. Soc. London **71**, 430 (1958)
5.13 A.E. Bowley, R.T. Delves, H.J. Goldsmid: Proc. Phys. Soc. London **72**, 401 (1958)
5.14 L.P. Caywood, R.G. Miller: Phys. Rev. B **2**, 3209 (1970)
5.15 L.R. Testardi, P.J. Stiles, E.H. Burstein: Bull. Am. Phys. Soc. **7**, 548 (1962)
5.16 R. Sehr, L.R. Testardi: J. Appl. Phys. **34**, 2754 (1963)
5.17 G.N. Ikonnikova, V.A. Kutasov, L.N. Luk'yanova, Sov. Phys. Solid State **32**, 1937 (1990)
5.18 H.J. Goldsmid: Proc. Phys. Soc. London **71**, 633 (1960)
5.19 R.B. Mallinson, J.R. Rayne, R.W. Ure: Phys. Rev. **175**, 1049 (1968)
5.20 J. Black, E.M. Conwell, L. Seigle, C.W. Spencer: J. Phys. Chem. Solids **2**, 240 (1957)
5.21 I.G. Austin: Proc. Phys. Soc. London **72**, 545 (1958)
5.22 D.L. Greenaway, G. Harbeke: J. Phys. Chem. Solids **26**, 1585 (1965)
5.23 S.K Mishara, S. Satpathy, O. Jepsen: J. Phys. Condens. Matter **9**, 461 (1997)
5.24 M. Bartkowiak, G. Mahan; private communication.
5.25 R. Mansfield, W. Williams: Proc. Phys. Soc. London **72**, 733 (1958)
5.26 H.J. Goldsmid, A.R. Sheard, D.A. Wright: Br. J. Appl. Phys. **9**, 365 (1958)
5.27 C.B. Satterthwaite, R.W. Ure: Phys. Rev. **108**, 1164 (1957)
5.28 H.J. Goldsmid: Proc. Phys. Soc. London Sect. B **69**, 203 (1956)
5.29 H.T. Langhammer, M. Stordeur, H. Sobotta, V. Riede: Phys. Status Solidi B **109**, 673 (1982)
5.30 M. Bartkowiak, G. Mahan: Proceedings, Eighteenth International Conference on Thermoelectrics, Baltimore (IEEE, Piscataway, New Jersey 1999) p. 713
5.31 F.D. Rosi, B. Abeles, R.V. Jensen: J. Phys. Chem. Solids **10**, 191 (1959)
5.32 H.J. Goldsmid: J. Appl. Phys. **32**, 2198 (1961)
5.33 U. Birkholz: Z. Naturforsch. Teil A **13**, 780 (1958)
5.34 C.H. Champness, W.B. Muir, P.T. Chiang: Can. J. Phys. **45**, 3611 (1967)
5.35 I.G. Austin, A.R. Sheard: J. Electron. Control **3**, 236 (1957)

5.36 S.V. Airapetyants, B.A. Efimova, T.S. Stavitskaya, L.S. Stil-bans, L.M. Sysoeva: Zh. Tekh. Fiz. **27,** 2167 (1957)
5.37 C.H. Champness, P.T. Chiang, P. Parekh: Can. J. Phys. **43,** 653 (1965)
5.38 W.M. Yim, F.D. Rosi: Solid State Electron. **15,** 1121 (1972)
5.39 M. Imamuddin, A. Dupre: Phys. Status Solidi A **10,** 415 (1972)
5.40 N.A. Bulatova, T.E. Svechnikova, S.N. Chizhevskaya: Inorg. Mater. **15,** 895 (1979)
5.41 N. Kh. Abrikosov, L.L. Ivanova: Inorg. Mater. **15,** 926 (1979)
5.42 K. Smirous, L. Stourac: Z. Naturforsch., Teil A, **14,** 848 (1959)
5.43 J.P. Fleurial, L. Gaillard, R.P. Triboulet: J. Phys. Chem. Solids **49,** 1237 (1988)
5.44 H.J. Goldsmid, J.W. Cochrane: Proceedings, Fourth International Thermoelectric Energy Conversion (IEEE, New York 1982)
5.45 W.M. Yim, A. Amith: Solid-State Electron. **15,** 1141 (1972)
5.46 N.Kh. Abrikosov, T.E. Svechnikova, S.N. Chizhevskaya SN: Inorg. Mater. **14,** 32 (1978)
5.47 A.L. Jain: Phys. Rev. **114,** 1518 (1959)
5.48 A.L. Jain, S.H. Koenig: Phys. Rev. **127,** 442 (1962)
5.49 B. Abeles, S. Meiboom: Phys. Rev. **101,** 544 (1956)
5.50 S.J. Freedman, H.J. Juretschke: Phys. Rev. **124,** 1379 (1961)
5.51 O. Öktü, G.A. Saunders: Proc. Phys. Soc., London **91,** 156 (1967)
5.52 C.G. Gallo, B.S. Chandrasekhar, P.H. Sutter: J. Appl. Phys. **34,** 144 (1963)
5.53 C. Uher, H.J. Goldsmid: Phys. Status Solidi B **65,** 765 (1974)
5.54 Ya. Korenblit, M.E. Kuznetsov, V.M. Muzhdaba, S.S. Shalyt: Sov. Phys. JETP **30,** 1009 (1969)
5.55 G.E. Smith, R. Wolfe, S.E. Haszko: Proceedings, International Conference on Physics of Semiconductors, Paris (Dunod, Paris 1964) p. 399
5.56 G.E. Smith, R. Wolfe: Proceedings, International Conference on Physics of Semiconductors, Kyoto (Dunod, Paris 1966) p. 651
5.57 R. Wolfe, G.E. Smith: Phys. Rev. **129,** 1086 (1963)
5.58 I.M. Tsidil'kovskii: *Thermomagnetic Effects in Semiconductors* (Infosearch, London, 1962)
5.59 C.F. Kooi: Report ASD-TDR-62-1100 Solid State Cryogenics, Final Report (1963)
5.60 A.W. Smith: Phys. Rev. **32,** 178 (1911)
5.61 B. Lenoir, A. Dauscher, M. Cassart, Y.I. Ravich, H. Scherrer: J. Phys. Chem. Solids **59,** 129 (1998)
5.62 E.G. Bowen, W. Morris Jones: Philos. Mag. **7,** 1029 (1932)
5.63 W.F. Ehret, M.B. Abramson: J. Am. Chem. Soc. **56,** 385 (1934)
5.64 G.E. Smith: Phys. Rev. Lett. **9,** 487 (1962)
5.65 N.B. Brandt, L.G. Lyubutina, N.A. Kryukova: Sov. Phys. JETP **26,** 93 (1968)
5.66 K.F. Cuff, R.B. Horst, J.L. Weaver, S.R. Hawkins, C.F. Kooi, G.W. Enslow: Appl. Phys. Lett. **2,** 145 (1963)

5.67 R.B. Horst, L.R. Williams: Proceedings, Third International Conference on Thermoelectric Energy Conversion (IEEE, New York, 1980) p. 139
5.68 G.E. Smith, R. Wolfe: J. Appl. Phys. **33,** 841 (1962)
5.69 R. Wolfe, G.E. Smith: Appl. Phys. Lett. **1,** 5 (1962)
5.70 P. Jandl, U. Birkholz, J. Appl. Phys. **76,** 7351 (1994)
5.71 M.E. Ertl, G.R. Pfister, H.J. Goldsmid: Br. J. Appl. Phys. **14,** 161 (1963)
5.72 C.B. Thomas, H.J. Goldsmid: Phys. Lett. A **27,** 369 (1968)
5.73 R.B. Horst, L.R. Williams: Proceedings, Third International Conference on Thermoelectric Energy Conversion, Arlington (IEEE, New York, 1980) p. 183
5.74 U. Birkholz: Proceedings, Eighth International Conference on Thermoelectric Energy Conversion (INPL, Nancy, France 1989) p. 98
5.75 A.F. Ioffe: *Semiconductor Thermoelements and Thermoelectric Cooling* (Infosearch, London 1957)
5.76 A.G. Ioffe, A.V. Airapetyants, A.V. Ioffe, N.V. Kolomoets, L.S. Stilibans: Dokl. Akad. Naur SSSR **106,** 981(1956)
5.77 V. Fano: In *CRC Handbook of Thermoelectrics*, Ed. by D.M. Rowe (CRC Press, Boca Raton 1995) p. 257
5.78 R.S. Allgaier, W.W. Scanlon: Phys. Rev. **111,** 1029 (1958)
5.79 D.M. Rowe, C.M. Bhandari: *Modern Thermoelectrics* (London, Holt, Rinehart & Winston, 1983)
5.80 D.A. Wright: Metall. Rev. **15,** 147 (1970)
5.81 F.D. Rosi, E.F. Hocking, N.E. Lindenblad: RCA Rev. **22,** 82 (1961)
5.82 E.A. Skrabek, D.S. Trimmer: In *CRC Handbook of Thermoelectrics*, Ed. by D.M. Rowe (CRC Press, Boca Raton 1995) p. 267
5.83 G.C. Christakudis, S.K. Plachkova, L.E. Shelimova, E.S. Avilov: Proceedings, Eighth International Conference on Thermoelectric Energy Conversion (INPL, Nancy, France, 1989) p. 125
5.84 C.B. Vining: In *CRC Handbook of Thermoelectrics*, Ed. by D.M. Rowe (CRC Press, Boca Raton 1995) p. 328
5.85 M.C. Steele, F.D. Rosi: J. Appl. Phys. **29,** 151 (1958)
5.86 D.M. Rowe, C.M. Bhandari: Appl. Energy **6,** 347 (1980)
5.87 G.A. Slack, M.A. Hussein: J. Appl. Phys. **70,** 2694 (1991)
5.88 G.J. Cosgrove, J.P. McHugh, W.A. Tiller: J. Appl. Phys. **32,** 621 (1961)
5.89 W.G. Pfann: *Zone Melting* (J. Wiley, New York, 1958)
5.90 W.A. Tiller, K.A. Jackson, J.W. Rutter, B. Chalmers: Acta Metal. **1,** 428 (1953)
5.91 D.T.J. Hurle: Solid-State Electron. **3,** 37 (1961)
5.92 M.J. Smith, R.J. Knight, C.W. Spencer: J. Appl. Phys. **33,** 2186 (1962)
5.93 J.P. McHugh, W.A. Tiller: Trans. Metall. Soc. AIME **215,** 651 (1959)
5.94 R.B. Horst, L.R. Williams: Proceedings, Fourth International Conference on Thermoelectric Energy Conversion, Arlington (IEEE, New York, 1982) p. 119
5.95 A.E. Vol'pyan, V.V. Marychev, V.V Shuyryaev, L.T. Evdokimenko: Inorg. Mater. **12** 1107 (1976)

5.96   A.C. Yang, F.D. Shepherd: J. Electrochem. Soc. **108**, 197 (1961)
5.97   F.J. Strieter: Adv. Energy Convers. **1**, 125 (1961)
5.98   E.M. Porbansky: J. Appl. Phys. **30**, 1455 (1959)
5.99   C. Uher: Ph.D. thesis, University of New South Wales (1975)
5.100 D.M. Brown, F.K. Heumann: J. Appl. Phys. **35**, 1947 (1964)
5.101 M.A. Short, J.J. Schott: J. Appl. Phys. **36**, 659 (1965)
5.102 R.G. Cope, A.W. Penn: J. Mater. Sci. **3**, 103 (1968)
5.103 R.A. Horne: J. Appl. Phys. **30**, 393 (1959)
5.104 T. Durst, H.J. Goldsmid, L.B. Harris: J. Mater. Sci. Lett. **16**, 2632 (1981)
5.105 S.V. Airapetyants, B.A. Efimova: Sov. Phys.-Tech. Phys. **3**, 1632 (1958)
5.106 H.J. Goldsmid, F.A. Underwood: Adv. Energy Convers. **7**, 297 (1968)
5.107 E.H. Volckmann, H.J. Goldsmid, J. Sharp: Proceedings, Fifteenth International Conference on Thermoelectric Energy Conversion, Pasadena (IEEE, New York 1996) p. 22
5.108 F.S. Samedov, M.M. Tagiev, D. S. Abdinov: Inorg. Mater. **34**, 704 (1998)
5.109 J. Seo, C. Lee, K. Park: Mater. Sci. Eng. B **54**, 135 (1998)
5.110 Y. Nakagira, H. Gyoten, Y. Yamamoto: Appl. Energy **59**, 147 (1998)
5.111 R.J. Buist: Proceedings, Third International Conference on Thermoelectric Energy Conversion, Arlington (IEEE, New York, 1980) p. 130
5.112 J.E. Parrott, A.W. Penn: Solid-State Electron. **3**, 91 (1961)
5.113 D. Ilzycer, A. Sher, M. Shiloh: Proceedings, Third International Conference on Thermoelectric Energy Conversion, Arlington (IEEE, New York 1980) p. 200
5.114 C.F. Kooi, R.B. Horst, K.F. Cuff: J. Appl. Phys. **39**, 4257 (1968)
5.115 B. Yates: J. Electron. Control **6**, 26 (1959)
5.116 D.K.C. Macdonald, E. Mooser, W.B. Pearson, I.M. Templeton, S.B. Woods: Philos. Mag. **4**, 433 (1959)
5.117 P.A. Walker: Proc. Phys. Soc. London **76**, 113 (1960)
5.118 H.J. Goldsmid: Proc. Phys. Soc. London **72**, 17 (1958)
5.119 H.J. Goldsmid, R.T. Delves: GEC J. **28**, 102 (1961)
5.120 M. Hansen: *Der Aufbau der Zweistofflegierung* (Springer, Berlin, Heidelberg 1936)

Chapter 6

6.1 G.A. Slack: In *CRC Handbook of Thermoelectrics*, Ed. by D.M. Rowe, (CRC Press, Boca Raton, 1995) p. 407
6.2 D.G. Cahill, S.K. Watson, R.O. Pohl: Phys. Rev. B **46**, 6131 (1992)
6.3 G.A. Slack: In *Solid State Physics*, Ed. by F. Seitz, D. Turnbull, H. Ehrenreich, (Academic Press, New York, 1979) pp. 1–71
6.4 G.A. Slack, V. Tsoukala: J. Appl. Phys. **76**, 1665 (1994)

6.5 See, for example, *Moffat's Handbook of Binary Phase Diagrams*, Vol. II (Genium Publ. Co., Schenectady, New York 1995) and references therein; T. Caillat, A. Borshchevsky, J.-P. Fleurial: J. Alloy Comp. **199**, 207 (1993)
6.6 See, for example, T. Caillat, J.-P. Fleurial, A. Borshchevsky: J. Crystl. Growth **166**, 722 (1996)
6.7 D.C. Johnson: Curr. Opin. Solid State Mater. Sci. **3**, 159 (1998)
6.8 M.D. Hornbostel, E.J. Hyer, J. Thiel, D.C. Johnson: J. Am. Chem. Soc. **119**, 2665 (1997)
6.9 M.D. Hornbostel, E.J. Hyer, J. Thiel, J.H. Evaldson, D.C. Johnson: Inorg. Chem. **36**, 4270 (1997)
6.10 H. Sellinschegg, S.L. Stuckmeyer, M.D. Hornbostel, D.C. Johnson: Chem. Mater. **10**, 1096 (1998)
6.11 G.S. Nolas, G.A. Slack, D.T. Morelli, T.M. Tritt, A.C. Ehrlich: J. Appl. Phys. **79**, 4002 (1996)
6.12 For details, see the recent review G.S. Nolas, D.T. Morelli, T.M. Tritt: Annu. Rev. Mater. Sci. **29**, 89 (1999)
6.13 C.B. Evers, L. Bonk, W. Jeitschko: Z. Anorg. Chem. **620**, 1028 (1994)
6.14 D.J. Braun, W. Jeitschko: J. Less Common Met. **72**, 147 (1980)
6.15 B.C. Sales, D. Mandrus, B.C. Chakoumakos, V. Keppens, J.R. Thompson: Phys. Rev. B **56**, 15081 (1997)
6.16 B.C. Chakoumakos, B.C. Sales, D. Mandrus, V. Keppens: Acta Cryst. B **5**, 341 (1999)
6.17 T.M. Tritt, G.S. Nolas, G.A. Slack, D.T. Morelli: J. Appl. Phys. **79**, 8412 (1996)
6.18 F. Hulliger: Helv. Phys. Acta **34**, 782 (1961)
6.19 R.N. Kuzimin, G.S. Zhdanov: Kristallografiya **5**, 869 (1960)
6.20 D.J. Singh, I.I. Mazin: Phys. Rev. B **56**, 1650 (1997)
6.21 H. Harima: J. Magn. Magn. Mater. **117**, 321 (1998)
6.22 G.S. Nolas, J.L. Cohn, G.A. Slack: Phys. Rev. B **58**, 164 (1998)
6.23 D.T. Morelli, G.P. Meisner, B. Chen, S. Hu, C. Uher: Phys. Rev. B **56**, 7376 (1997)
6.24 B.C. Sales, B.C. Chakoumakos, D. Mandrus: Phys. Rev. B **61**, 2475 (2000)
6.25 G.P. Meisner, D.T. Morelli, S. Hu, J. Yang, C. Uher: Phys. Rev. Lett. **80**, 3551 (1998)
6.26 G.S. Nolas, V.G. Harris, T.M. Tritt. G.A. Slack: J. Appl. Phys. **80**, 6304 (1996)
6.27 T. Caillat, J. Kulleck. A. Borshchevsky, J.-P. Fleurial: J. Appl. Phys. **80**, 8419 (1996)
6.28 L.D. Dudkin, N.K. Abrikosov: Zh. Neorg. Khim. **1**, 2096 (1956)
6.29 D.J. Singh, W.J. Pickett: Phys. Rev. B **50**, 11235 (1994)
6.30 T. Caillat, J.-P. Fleurial, A. Borshchevsky: J. Crystl. Growth **166**, 722 (1996)
6.31 J.W. Sharp, E.C. Jones, R.K. Williams, P.M. Martin, B.C. Sales: J. Appl. Phys. **78**, 1013 (1995)

6.33 N.R. Dilley, E.J. Freeman, E.D. Bauer, M.B. Maple: Phys. Rev. B **58**, 6287 (1998)
6.34 A. Leither-Jasper, D. Kaczorowski, P. Rogl, J. Bogner, M. Reissner, W. Steiner, G. Wiesinger, C. Godart: Phys. Rev. B **109**, 395 (1999)
6.35 G.S. Nolas, M. Kaeser, T.M. Tritt, H. Sellinschegg, D.C. Johnson: Materials Research Society Symposium Proceedings, in press.
6.36 See, for example, F. Franks: *Water, A Comprehensive Treatise* (Plenum Press, New York 1973)
6.37 C. Cros, M. Pouchard, P. Hagenmuller: C.R. Acad. Sc. Paris 260, 4764 (1965); J.S. Kasper, P. Hagenmuller, M. Pouchard, C. Cros: Science **150**, 1713 (1965); C. Cros, M. Pouchard, P. Hagenmuller, J.S. Kasper: Bull. Soc. Chim. Fr. **7**, 2737 (1968)
6.38 J. Gallmeier, H. Schäffer, A. Weiss: Zeit. Naturforschg. **24B**, 665 (1969)
6.39 G.S. Nolas, J.L. Cohn, G.A. Slack, S.B. Schujman: Appl. Phys. Lett. **73**, 178 (1998) and G.S. Nolas, G.A. Slack, J.L. Cohn, S.B. Schujman: Proceedings, Seventeenth International Conference on Thermoelectrics, Baltimore (IEEE, Piscataway, New Jersey 1998) p. 294
6.40 See, for example, D.G. Cahill, S.K. Watson, R.O. Pohl: Phys. Rev. B **46**, 6131 (1992) and references therein
6.41 H. Kawaji, H. Horie, S. Yamanaka, M. Ishikawa: Phys. Rev. Lett. **74**, 1427 (1995)
6.42 J.D. Bryan, V.I. Srdanov, G. Stucky, D. Schmidt: Phys. Rev. B **60**, 3064 (1999)
6.43 G.B. Adams, M. O'Keeffe, A.A. Demkov, O.F. Sankey, Y.-M. Huang: Phys. Rev. B **49**, 8048 (1994)
6.44 B. Einsenmann, H. Schäfer, J. Zagler: J. Less Common Metals **118**, 43 (1986)
6.45 H.G. von Schnering, W. Carrillo-Cabrera, R. Kröner, E.-M. Peters, K. Peters, R. Nesper: Z. Kristallogr. **213**, 679 (1998)
6.46 G.S. Nolas: Mater. Res. Soc. Symp. Proc. **545**, 435 (1999)
6.47 G. Ramachandran, J. Dong, J. Diefenbacher, J. Gryko, R. Marzke, O. Sankey, P. McMillan: J. Solid State Chem. **145**, 716 (1999)
6.48 E. Reny, P. Gravereau, C. Cros, M. Pouchard: J. Mater. Chem. **8**, 2839 (1998)
6.49 G.S. Nolas, T.J.R. Weakley, J.L. Cohn, R. Sharma: Phys. Rev. B **61**, 3845 (2000)
6.50 S.B. Schujman, G.S. Nolas, R.A. Young, C. Lind, A.P. Wilkinson, G.A. Slack, R. Patschke, M.G. Kanatzidis, M. Ulutagay, S.-J. Hwu: J. Appl. Phys. **87**, 1529 (2000)
6.51 B.C. Chakoumakos, B.C. Sales, D.G. Mandrus, G.S. Nolas: J. Alloys Comp. **296**, 801 (1999)
6.52 T.L. Chu, S.S. Chu, R.L. Ray: J. Appl. Phys. **53**, 7102 (1982)
6.53 G.S. Nolas, T.J.R. Weakley, J.L. Cohn: Chem. Mater. **11**, 2470 (1999)

6.54 G.S. Nolas, B.C. Chakoumakos, B. Mahieu, G.J. Long, T.J.R. Weakley: Chem. Mater., in press
6.55 J.L. Cohn, G.S. Nolas, V. Fessatidis, T.H. Metcalf, G.A. Slack: Phys. Rev. Lett. **82**, 779 (1999)
6.56 B.B. Iverson, A.E.C. Palmqvist, D.E. Cox, G.S. Nolas, G.D. Stucky, N.P. Blake, H. Metiu: J. Solid State Chem. **149**, 455 (1999)
6.57 R. Nesper, K. Vogel, P. Blochl: Angew. Chem. **32**, 701 (1993)
6.58 W. Sekkal, S. Ait Abderahmane, R. Terki, M. Certier, H. Aourag: Mater. Sci. Eng. **B64**, 123 (1999)
6.59 M. Menon, E. Richter, K.R. Subbaswamy: Phys. Rev. B **56**, 12290 (1997)
6.60 D. Kahn, J.P. Lu: Phys. Rev. B **56**, 13898 (1997)
6.61 J. Dong, O.F. Sankey, G. Kern: Phys. Rev. B **60**, 950 (1999)
6.62 A.A. Demkov, O.F. Sankey, K.E. Schmidt, G.B. Adams, M. O'Keeffe: Phys. Rev. B **50**, 17001 (1995)
6.63 S. Saito, A. Oshiyama: Phys. Rev. B **51**, 2628 (1995)
6.64 N.P. Blake, L. Mollnitz, G. Kresse, H. Metiu: J. Chem. Phys. **111**, 3133 (1999)
6.65 J. Zhao, A. Buldum, J.P. Lu, C.Y. Fong: Phys. Rev. B **60**, 14177 (1999)
6.66 J. Dong, O.F. Sankey: J. Phys. Condens. Matter. **11**, 6129 (1999)
6.67 G.S. Nolas, J.L. Cohn, B.C. Chakoumakos, G.A. Slack: Proceedings, Twenty Fifth International Thermal Conductivity Conference (Technomic, Lancaster, Pennsylvania, 2000) p. 122
6.68 B.C. Sales, B.C. Chakoumakos, D. Mandrus, J.W. Sharp: J. Solid State Chem. **146**, 528 (1999)
6.69 G.S. Nolas, J.L. Cohn, E. Nelson: Proceedings, Nineteenth International Conference on Thermoelectrics, Baltimore (IEEE, Piscataway, New Jersey 1999) p. 493
6.70 C. Cros, M. Pouchard, P. Hagenmuller: J. Solid State Chem. **2**, 570 (1970)
6.71 N.F. Mott: J. Solid State Chem. **6**, 348 (1973)
6.72 G.S. Nolas, unpublished results

Chapter 7

7.1 D. Elwell, H.J. Scheel: *Crystal Growth from High-Temperature Solutions* (Academic Press, New York 1975); H.J. Scheel: J. Crystl. Growth **24/25**, 669 (1974); R. Sanjines, H. Berger, F. Levy: Mater. Res. Bull. **23**, 549 (1998); R.W. Garner, W.B. White: J. Crystl. Growth, **7**, 343 (1970)
7.2 M.G. Kanatzidis, T.J. McCarthy, T.A. Tanzer, L. Chen, L. Iordanidis, T. Hogan, C.R. Kannewurf, C. Uher, B. Chen: Chem. Mater. **8**, 1465 (1996)
7.3 B. Chen, C. Uher, L. Iordanidis, M.G. Kanatzidis: Chem. Mater. **9**, 1655 (1997)

7.4 D. Chung, K. Choi, L. Iordanidis, J.L. Schindler, P.W. Brazis, C.R. Kannewurf, B. Chen, S. Ju, C. Uher, M.G. Kanatzidis: Chem. Mater. **9**, 3060 (1997)
7.5 D. Chung, L. Iordanidis, K. Choi, M.G. Kanatzidis: Bull. Korean Chem. Soc. **19**, 1283 (1999)
7.6 M.G. Kanatzidis, D. Chung, L. Iordanidis, K. Choi, P. Brazis, M. Rocci, T. Hogan, C.R. Kannewurf: Mater. Res. Soc. Symp. Proc. **545**, 233 (1999)
7.7 D. Chung, T. Hogan, P. Brazis, M. Rocci-Lane , C.R. Kannewurf, M. Bastea, C. Uher, M.G. Kanatzidis: Science **287**, 1024 (2000)
7.8 S.A. Sunshine, D. Kang, J.A. Ibers: J. Am. Chem. Soc. **109**, 6202 (1987) 6204; M.A. Pell, J.A. Ibers: Chem. Ber. **129**, 1207 (1996)
7.9 W. Bensch, P. Durichen: Chem. Ber., **129**, 1207 (1996)
7.10 H.J. Goldsmid, J.W. Sharp: J. Electron. Mater. **28**, 869 (1999)
7.11 P. Larson, S.D. Mahanti, D. Chung, M.G. Kanatzidis: unpublished results
7.12 M. Izumi, K. Uchinokura, E. Matsuura: Solid State Commun. **37**, 641 (1981)
7.13 S. Okada, T. Sambongi, M. Ido: J. Phys. Soc. Jpn. **49**, 839 (1980)
7.14 T.J. Wieting, D.U. Gubser, S.A. Wolf, F. Levy: Bull. Am. Phys. Soc. **25**, 340 (1980)
7.15 F.J. DiSalvo, R. Fleming, J. Waszczak: Phys. Rev. B. **24**, 2935-9 (1981)
7.16 Okada, Sambongi, Ido, Tazuke, Aoki, Fujita: J. Phys. of Jpn. **51**, 460 (1982)
7.17 Furuseth, L. Brattas, A. Kjeskshus: Acta Chem. Scand. **27**, 2367 (1973).
7.18 Brattas, A. Kjekshus: Acta Chem. Scand. **25**, 2783 (1971).
7.19 N. Kamm, D.J. Gillespie, A.C. Ehrlich, T.J. Wieting, F. Levy: Phys. Rev. B. **31**, 7617 (1985).
7.20 N. Kamm, D.J. Gillespie, A.C. Ehrlich, D.L. Peebles, F. Levy: Phys. Rev. B. **35**, 1223 (1987).
7.21 J. Isumi, T. Nakayama, K. Uchinokura, S. Harada, R. Yoshisaki, E. Matsuura: J. Phys. C **20**, 3691 (1987)
7.22 M. Whangbo, F.J. Disalvo, R. Fleming: Phys. Rev. B. **26**, 687 (1982)
7.23 M. Izumi, K. Uchinokura, E. Matsura, S. Harada: Solid State Commun. **42**, 773 (1982)
7.24 T.M. Tritt, N. Lowhorn, R. Littleton, A. Pope, C. Feger, J. Kolis: Phys. Rev. B. **60**, 7816 (1999)
7.25 T.M. Tritt, M. Wilson, R. Littleton, C. Feger, J. Kolis, A. Johnson, D. Verebelyi, S. Hwu, M. Fakhruddin, F. Levy: Mater. Res. Soc. Proc. **478**, 249 (1997)
7.26 R.T. Littleton IV, T.M. Tritt, C.R. Feger, J.W. Kolis: Mater. Res. Soc. Proc. **545,** 381 (1998)
7.27 T.E. Jones, W.W. Fuller, T.J. Wieting, F. Levy: Solid State Commun. **42**, 793 (1982)
7.28 W.W. Fuller, S. Wolf, T. Weiting, R. LaCoe, P. Chaikin, C. Huang: J. Phys. **C3**, 1709 (1983)

7.29 R.T. Littleton IV, T.M. Tritt, C.R. Feger, J. Kolis, M.L. Wilson, M. Marone, J. Payne, D. Verebeli, F. Levy: Appl. Phys. L. **72**, 2056 (1998)

7.30 E.P. Stillwell, A.C. Ehrlich, G.N. Kamm, D.J. Gillespie: Phys. Rev. B. **39**, 1626 (1989)

7.31 W.M. Yim, F.D. Rosi: Solid State Electron. **15**, 1121 (1972)

7.32 R.T. Littleton IV, T.M. Tritt, J.W. Kolis, D. Ketchum: Phys. Rev. B. **60**, 13453 (1999)

7.33 E. Dichi, G. Kra, R. Eholié, B. Legendre: J. Alloys Comp. **194**, 147 (1993); **194**, 155 (1993); **199**, 7 (1993); **199**, 21 (1993)

7.34 A. Abba-Touré, G. Kra, R. Eholié: J. Less Common Met. **170**, 199 (1991)

7.35 V. Agafanov, B. Legendre, N. Rodier, J. M. Cense, E. Dichi, G. Kra: Acta Cryst. C **47**, 850 (1991)

7.36 J.W. Sharp, B.C. Sales, D.G. Mandrus and B.C. Chakoumakos: Appl. Phys. Lett. **74**, 3794 (1999)

7.37 B.C. Sales, B.C. Chakoumakos, D. Mandrus, J.W. Sharp: J. Solid State Chem. **146**, 528 (1999)

7.38 B. Eisenmann, H. Schwerer, H. Schäfer: Mater. Res. Bull. **18**, 383 (1983)

7.39 C. Brinkmann, B. Eisenmann, H. Schäfer: Mater. Res. Bull. **20**, 299 (1985)

7.40 R.E. Marsh: J. Solid State Chem. **87**, 467 (1990)

Chapter 8

8.1 H.B.G. Casimir, Physica **5**, 495 (1938)

8.2 H.J. Goldsmid, A. W. Penn: Phys. Lett. **27A**, 523 (1968)

8.3 H.J. Goldsmid: High Temperatures – High Pressures **1**, 153 (1969)

8.4 J.E. Parrott: J. Phys. C: Solid State. Phys. **2**, 147 (1969)

8.5 N. Savvides, H.J. Goldsmid: J. Phys. C: Solid State Phys. **13**, 4657 (1980)

8.6 Y.F. Komnik, E.I. Bukhstab: Sov. Phys. JETP **27**, 34 (1968)

8.7 D.D. Thornburg, C.M. Wayman: Philos. Mag. **20**, 1153 (1969)

8.8 L.D. Hicks, M.S. Dresselhaus: Phys. Rev. B **47**, 12727 (1993)

8.9 D.A. Broido, T.L. Reinecke: Phys. Rev. B **51**, 13797 (1995)

8.10 L.D. Hicks, M.S. Dresselhaus: Phys. Rev. B **47**, 16631 (1993)

8.11 D.A. Broido, T.L. Reinecke: Appl. Phys. Lett. **67**, 100 (1995)

8.12 L.D. Hicks, T.C. Harman , M.S. Dresselhaus: Appl. Phys. Lett. **63**, 3230 (1993)

8.13 T. Koga, X. Sun, S.B. Cronin, M.S. Dresselhaus: Appl. Phys. Lett. **75**, 2438 (1999)

8.14 T. Koga, T.C. Harman, X. Sun, S.B. Cronin, M.S. Dresselhaus: Mater. Res. Soc. Symp. Proc. **545**, 479 (1999)

8.15 S.Y. Ren, J.D. Dow: Phys. Rev. B **25**, 3750 (1981)

8.16 G. Chen, M. Neagu: Appl. Phys. Lett. **71**, 2761 (1997)

8.17 G. Chen: Phys. Rev. B **57**, 14958 (1997)

8.18 S.-M. Lee, D. Cahill, R. Venkatasubramanian: Appl. Phys. Lett. **70**, 2957 (1997)
8.19 T. Yao: Appl. Phys. Lett. **51**, 1798 (1987)
8.20 T.H. Geballe, G.W. Hull: Conf. De Phys. des Basses Temps. # 86, (Paris, 1955)
8.21 C. Herring: Phys. Rev. **96**, 1163 (1954)
8.22 P. Hyldgaard, G.D. Mahan: Phys. Rev. B **56**, 10754 (1997)
8.23 M. Bartkowiak, G.D. Mahan: Mater. Res. Soc. Symp. Proc. **545**, 265 (1999)
8.24 T.E. Whall, E.H.C. Parker: Proceedings, First European Conference on Thermoelectrics (Peter Peregrinus, London, 1987) p. 51
8.25 G. Abstreiter, H. Brugger, T. Wolf, H. Jorke, H.J. Herzog: Phys. Rev. Lett. **54**, 2441 (1985)
8.26 R. People, J.C. Bean: Appl. Phys. Lett. **48**, 538 (1986)
8.27 T.C. Harman, D.L. Spears, M.J. Manfra: J. Elec. Mater. **25**, 1121 (1996)
8.28 R. Venkatasubramanian, T. Colpitts, B.O'Quinn, S. Liu, N. El-Masry, M. Lamvik: Appl. Phys. Lett. **75**, 1104 (1999)
8.29 I. Yamasaki, R. Yamanaka, M. Mikami, H. Sonobe, Y. Mori, T. Sasaki: Proceedings, Seventeenth International Conference on Thermoelectrics, Baltimore (IEEE, Piscataway, New Jersey, 1998) p. 210
8.30 R. Venkatasubramanian: Naval Res. Rev. **58**, 44 (1998)
8.31 S.-M. Lee, D.G. Cahill, R. Venkatasubramanian: Appl. Phys. Lett. **70**, 2957 (1997)
8.32 T. Borca-Tasciuc, D. Song, J.L. Liu, G. Chen, K.L. Wang, X. Sun, M.S. Dresselhaus, T. Radetic, R. Gronsky: Mater. Res. Soc. Symp. Proc. **545**, 473 (1999)
8.33 Y. Okamoto, H. Uchino, T. Kawahara, J. Morimoto: Jpn. J. Appl. Phys. **38**, L945 (1999)
8.34 T.C. Harman, D.L. Spears, M.P. Walsh: J. Electron. Mater. Lett. **28**, L1 (1999)
8.35 T. Koga, T.C. Harman, X. Sun, S.B. Cronin, M.S. Dresselhaus: Mater. Res. Soc. Symp. Proc. **545**, 479 (1999)
8.36 E.G. Bauer, B.W. Dodson, D. J. Ehrlich, L.C. Feldman, C.P. Flynn, M.W. Geis, J.P. Harbison, R.J. Matyi, P.S. Peercy, P.M. Petroff, J.M. Phillips, G. B. Stringfellow, A. Zangwill: J. Mater. Res. **5**, 852 (1990)
8.37 T.C. Harman, P.J. Taylor, D.L. Spears, M.P. Walsh: Proceedings, Eighteenth International Conference on Thermoelectrics, Baltimore (IEEE, Piscataway, New Jersey, 1999) p. 280
8.38 J. Cibert, P.M. Petroff, G.J. Dolan, S.J. Pearton, A.C. Gossard, J.H. English: Appl. Phys. Lett. **49**, 1275 (1986)
8.39 Z. Zhang, J.Y. Ying, M.S. Dresselhaus: J. Mater. Res. **13**, 1745 (1998)
8.40 T.E. Huber, M.J. Graf, C.A. Foss: Mater. Res. Soc. Symp. Proc. **545**, 227 (1999)

8.41 D.L. Demske, J.L. Price, N.A. Guardala, N. Lindsey, J.H. Barkyoumb, J. Sharma, H.H. Kang, L. Salamanca-Riba: Mater. Res. Soc. Symp. Proc. **545**, 209 (1999)
8.42 Dmitri Routkevitch (Nanomaterials Research Corp.); private communication
8.43 J. Heremans, C.M. Thrush, Z. Zhang, X. Sun, M.S. Dresselhaus, J.Y. Ying, D.T. Morelli: Phys. Rev. B **58**, R10091 (1998)
8.44 S.B. Cronin, Y.M. Lin, T. Koga, X. Sun, J.Y. Ying, M.S. Dresselhaus: Proceedings, Eighteenth International Conference on Thermoelectrics, Baltimore (IEEE, Piscataway, New Jersey, 1999) p. 554
8.45 Z. Zhang, X. Sun. M.S. Dresselhaus, J.Y. Ying, J.P. Heremans: Appl. Phys. Lett. **73**, 1589 (1998)
8.46 G. Min, D.M. Rowe, F. Volklein: Electron. Lett. **34**, 222 (1998)
8.47 F. Völklein, G. Min, D.M. Rowe: Sensors and Actuators **75**, 95 (1999)
8.48 A.R. Ubbelohde: *Intercalated Layered Materials*, Ed. by F.A. Lévy (Reidel, Dordrecht, 1979)
8.49 S.B. Cronin, T. Koga, X. Sun, Z. Ding, S.-C. Huang, R.B. Kaner, M.S. Dresselhaus: Mater. Res. Soc. Symp. Proc. **545**, 397 (1999)
8.50 M.R. Agiar, R. Caram: J. Crystl. Growth **175**, 70 (1997)

Chapter 9

9.1 G.D. Mahan, J.O. Sofo, M. Bartkowiak, J. Appl. Phys. **83**, 4683 (1998)
9.2 A.F. Ioffe: *Semiconductor Thermoelements and Thermoelectric Cooling* (Infosearch, London, 1957)
9.3 G.D. Mahan: J. Appl. Phys. **76**, 4362 (1994)
9.4 J.L Dye: Chemtracts: Inorg. Chem. **5**, 243 (1993)
9.5 R.H. Huang, J.L. Dye: Chem. Phys. Lett. **166**, 133 (1990)
9.6 G.S. Nolas, H.J. Goldsmid: J. Appl. Phys. **85**, 4066 (1999)
9.7 A. Shakouri, C. LaBounty, J. Piprek, P. Abraham, J.E. Bowers: Appl. Phys. Lett. **74**, 88 (1999)
9.8 C.B. Vining, G.D. Mahan: Appl. Phys. Lett. **86**, 6852 (1999).
9.9 R.P. Chasmar, R. Stratton: J. Electron. Control **7**, 52 (1959)
9.10 G.Chen: Phys. Rev. B **57**, 14958 (1998)

Subject Index

3ω method 98–99, 245

Acceptor impurites 33, 59
Altenkirch 1
Anisotropic
– materials 5, 75, 92
– Peltier effect 84–85
– Seebeck effect 84–85
Antimony 131–132
Antimony telluride 122
Artificially structured materials 235
Atomic displacement parameters 182, 199–201, 203, 210, 230

Ballistic transport 255
Band gap 31, 33, 65, 66, 68, 70, 71, 75, 177
– layered materials 238
– silicon 31
Band structure 30
– bismuth telluride 113–116
– complex 74–75
– layered materials 238
– multivalley 74–75
– nonparabolic 51–55, 75
– parabolic 36
– silicon 31–32
Barrier layers 238
Barrier thermal resistance 264
Beta parameter 59–62, 70, 88, 89, 236, 268

Bi–Sb 12, 128, 129, 137–146
– as thermoelectric material 140–142, 173
– as thermomagnetic material 143–145, 176
– band structure 137–138
– carrier mobilities 139
– effective mass 138
– energy gap 137–139
– lattice thermal conductivity 139–140
– melt growth 157–159
– powder metallurgy 161–162
– Seebeck coefficient 140–141
Bipolar conduction 42–44, 65, 67
Bismuth 87, 131
– as thermoelectric material 132–134
– as thermomagnetic material 135–137, 144
– thin layers 238
Bismuth selenide 121–122
Bismuth telluride 111–121
– anisotropy 121
– band structure 113–115
– Brillouin zone 114–115
– carrier concentration 117–118
– crystal structure 111–112
– diffusion in 111–112
– dopants 119–120
– effective mass 116, 118–119, 121
– electrical conductivity 114, 117
– energy gap 116, 120
Bismuth telluride (cont.)
– figure of merit 1, 121
– films 162

288   Subject Index

– general properties 111–112
– melt growth 155–157
– n-type alloys 123–128, 173
– p-type alloys 86, 123–128
– powder metallurgy 159–162
– scattering parameter 113, 118
– Seebeck coefficient 116, 118–119 –
 standard materials 129–131, 229
– superlattices 244–245
– thermal conductivity 120, 121
– transport properties 116–121
Bonding
– covalent 16, 191
– ionic 15, 72
– metallic 17
– van der Waals 17
Bose–Einstein distribution 25
Bridgman relationship 7
Brillouin zone
– bismuth telluride 114, 115
– electrons 30
– fcc lattice 31
– phonons 22
Bulk modulus 81

Carnot 2
– efficiency 10, 12, 262, 265, 266
Carrier concentration 40, 53, 59, 87
Clathrate hydrates 191
Clathrate semiconductors 191–207
– band structure 198
– bonding 196–197
– carrier mobility 205
– crystal structures 191–193, 195–198
– electronic properties 205–207
– energy gap 193, 198
– lattice thermal conductivity 192, 194, 202–204
Clathrate semiconductors (cont.)
– synthesis 194, 195
$CoAs_3$ 178
Coefficient of performance 10–11
– module 168–172
– thermionic 256, 258–261, 267

Combinatorial synthesis 180, 211
Compressibility 76, 77, 80, 81
Conduction band 33, 34, 36, 71
Constitutional supercooling 153
Contact resistance 91, 104, 167, 170, 253
CoP_3 181
$CoSb_3$ 178
$CsBi_4Te_6$ 216–220
– band structure 220
– carrier mobilities 218, 230
– crystal structure 217, 218
– electrical conductivity 217–219
– figure of merit 218–220
– Seebeck coefficient 217–219
– thermal conductivity 217
Cs_8Sn_{44} 203, 204

De Broglie wavelength 236, 238
Debye 18
– frequency 19
– lattice model 18, 26
– specific heat 19, 20
– temperature 19, 76, 77, 88
Deformation potential 71, 238
Degenerate electron gas 30, 41, 54
Density of states 28, 32, 209, 220
– effective 41, 60, 236
Devices (see Modules)
Diamond 18
– structure 15–17, 191
Dielectric constant 72, 73
Donor impurities 33, 59
Dulong–Petit law 18

Effective mass 32, 38–39, 71, 74, 75, 237–238
– anisotropic 75
– nonparabolic 52
Electrical conductivity 5, 239
– in magnetic field 50
– multiband 43, 65, 68
– nonparabolic band 53
– semiconductor 37, 39, 40, 42, 61, 62

Electrical resistance
– of couple 9
Electrical resistivity 5, 8, 9, 235
– adiabatic 13, 47
– isothermal 47, 68
– measurement 91–92
Electronegativity 17, 177
Electrons 28, 36, 70
– alloy scattering 35, 73
– equilibrium with phonons 242
– impurity scattering 35, 63, 70, 73, 75, 88
– intervalley scattering 75
– lattice scattering 35, 63, 71, 72
– scattering 34, 63
Energy bands (see Band structure)
Energy gap (see Band gap)
Ettingshausen
– coefficient 6, 13, 14
– cooling 13, 48, 49, 105, 107
$Eu_8Ga_{16}Ge_{30}$ 202, 203

Fermi–Dirac distribution 28, 29
Fermi–Dirac integrals 39–41, 61, 237
Fermi energy 29, 33, 220, 236–238 –
  reduced 39, 59–67, 69, 89
Figure of merit 10, 91, 177, 238, 261
– anisotropic element 86–88
– couple 11
– device 104–105
Figure of merit (cont.)
– effective 166
– measurement 99–105
– multiband 65, 67
– optimized 59, 61, 63, 64
– phonon drag 87, 88
– very low temperature 88, 89
Flux growth 211–212

Generation efficiency 12
Grain size 235
Grüneisen parameter 76–78, 82

Hall coefficient 6, 48–50, 68
– nonparabolic band 53
Harman's method 102–106
Heat conduction
– by lattice 18
– losses 96, 111–113, 164–167, 251, 252, 257
Heterostructural devices 263
$HfTe_5$ 220–226, 227
– anisotropy 222
– band structure 223
– crystal structure 221, 222
– electrical resistivity 221, 223–228
– magnetoresistance 223
– phonon drag 224
– pressure effects 225
– Seebeck coefficient 223–228
– solid solutions 223–229
– synthesis 222
– solid solutions 223–229
– thermal conductivity 229
Holes 32, 36, 70

Intercalation 253
Ioffe 79, 146, 255
Iron disilicide 146
$IrSb_3$ 181

Joule heat 9, 99, 103

$\beta\text{-}K_2Bi_8Se_{13}$ 212–216
– energy gap 213
– thermal conductivity 214
$K_{2.5}Bi_{8.5}Se_{14}$ 212–216
– energy gap 213
– thermal conductivity 214
K_8Ge_{46} 198
Kelvin 2
– relationships 4, 5

Lead chalcogenides 72
Lorenz number 41, 42, 61, 62, 239, 242

Mass density 77, 78, 80, 241
Maxwell–Boltzmann distribution 28, 30
Metastable synthesis 180
Mixed valence (see Valence fluctuation)
Mobility 40–41, 60, 61, 63, 67, 68, 70, 71
– anisotropic 75
– Hall 50
– low temperature 89
– nonparabolic band 52, 53
Modulated element reactants 180
Modules
– multistage 10, 170–173
– single-stage 163–170
– thermomagetic 173–175
Multivalley bands 31, 35, 74, 74, 177

Nanowires 238, 248–251
– electrical resistivity 248
– magnetoresistance 250
Nernst coefficient 6, 13–14, 47, 50, 67, 107
– nonparabolic band 54
Nondegenerate electron gas 30, 39, 59, 60, 67

PbTe 75, 146–147
– superlattices 243, 247, 248
Peltier 2
– anisotropic effect 84, 85
– coefficient 2, 38, 56
– heat 8, 91, 92, 99, 105, 107
Pentatelluride compounds (see $HfTe_5$ and $ZrTe_5$)
PGEC 176, 177, 207, 210
Phonon drag 36, 56, 57, 87
Phonons 21
– boundary scattering 23–24, 83, 84, 235, 236, 240, 241
– defect scattering 23–28, 80, 81, 235
– dispersion 241

– equilibrium with electrons 242
– mean free path 22, 23, 76, 83, 84, 89, 236
– normal scattering 22, 80
– Umklapp scattering 22–23, 80, 235, 239
Poisson's ratio 81

Quantum wells 237, 238, 242–244, 251
Quantum wires 238

Rattling motion 178, 179, 182, 184, 186, 189, 191, 196, 199–201, 210, 214, 217
Relaxation time
– electron 34–35, 71–73
– phonon 24, 26–27, 87, 241–242
Richardson equation 256
Righi–Ludec 7, 107

Scattering parameter 35, 38, 41, 60–66, 71–73, 114, 236, 237, 239
Seebeck 2
– anisotropic effect 84–87
– effect 91, 99, 107
Seebeck coefficient 2, 8–9, 38–39, 41, 42, 99, 235, 239
– in magnetic field 47
– local 94–95
– low temperature 87
– measurement 93–94
– multiband 43, 65–66
– nonparabolic band 54–55
– optimized 61, 62, 64, 66
– phonon drag 56, 57, 241
– thin films
Segregation 152–154
Semiconductor 15, 33, 71
– extrinsic 33–34, 48, 67
– intrinsic 33–34, 48, 67
– nonparabolic band 51
– parabolic band 36
Semimetal 33, 48, 69, 71

Si–Ge 80, 84, 146, 149, 151, 235
- band structure 150
- superlattice 232–242, 245–246
Size effects 235–236
Skutterudite
- band structure 183
- binary 178–181, 183, 184
- bonding 183
- crystal structure 181
- electronic properties 187–189
- electronic thermal conductivity 184, 186
- energy gap 183
- figure of merit 189–191
- filled 181–191, 202
- lattice thermal conductivity 184–187
- partially filled 189
Space charge 258–259
Specific heat
- electronic 28, 29
- lattice 19, 22, 88, 236, 240
Speed of sound 18–19, 76, 81, 89, 236
$Sr_8Ga_{16}Ge_{30}$ 192, 194, 195, 199–204, 206
Stefan–Boltzmann constant 257, 258
Strain 238
Superconductors
- passive legs 12, 87
Superlattices
- Bi_2Te_3/Sb_2Te_3 244, 245
- GaAs/AlAs 245, 247
- lattice thermal conductivity 244
- PbTe-based 243, 247, 248
- Si/Ge 239–242, 245, 246

TAGS 147–149
Thermal conductance
- of couple 9
- barrier 264
Thermal conductivity 5, 8–9, 51, 65, 91, 235
- measurement 95–97

Thermal conductivity, electronic 38–39, 239
- bipolar 44, 51
- in magnetic field 47, 51
Thermal conductivity, lattice 39, 60, 65, 67, 70, 76–79, 268
- Eucken's rule 21, 23
- kinetic formula 22
- low temperature 88, 89, 127, 128, 241
- minimum 178
- solid solutions 79–82
- superlattices 240, 244
- various semiconductors 78, 79
- with point defects 27, 80
Thermal diffusivity 97
- measurement 97–99
Thermal expansion coefficient 76, 179
Thermionic generation 255
Thermionic refrigeration 255, 263
Thermionic transport 254, 263, 264, 269
Thermomagnetic
- cascade 174–175
- coefficients 5–7, 107, 108
- device 70, 173–175
- effects 5–7
- figure of merit 13–14, 67–70, 108–109
- measurements 105–109
Thomson coefficient 2, 4, 9
Tl_2GeTe_5 229–233
- carrier mobility 231
- crystal structure 229–231
- electrical resistivity 231, 233
- Seebeck coefficient 231, 232
- synthesis 230
- thermal conductivity 232, 233
Tl_2SnTe_5 229–233
- carrier mobility 231
- crystal structure 229–231
- electrical resistivity 231, 233
- Seebeck coefficient 231–232

– synthesis 230
– thermal conductivity 232–233
Tunneling effects 238, 263

Valence band 33–34, 36, 71
Valence fluctuation 187

Wiedemann–Franz–Lorenz law 42, 88, 120, 258
Work function, 255–257, 259, 261–264, 266–267

$ZrTe_5$ 220–226, 228
– anisotropy 222
– band structure 223
– crystal structure 221, 222
– electrical resistivity 221, 223–228
– phonon drag 224
– pressure effects 226
– Seebeck coefficient 223–228
– solid solutions 223–229
– synthesis 222
– thermal conductivity 229

Springer Series in
MATERIALS SCIENCE

Editors: R. Hull   R. M. Osgood, Jr.   H. Sakaki   A. Zunger

1 Chemical Processing with Lasers*
   By D. Bäuerle

2 Laser-Beam Interactions with Materials
   Physical Principles and Applications
   By M. von Allmen and A. Blatter
   2nd Edition

3 Laser Processing of Thin Films
   and Microstructures
   Oxidation, Deposition and Etching
   of Insulators
   By. I. W. Boyd

4 Microclusters
   Editors: S. Sugano, Y. Nishina, and S. Ohnishi

5 Graphite Fibers and Filaments
   By M. S. Dresselhaus, G. Dresselhaus,
   K. Sugihara, I. L. Spain, and H. A. Goldberg

6 Elemental and Molecular Clusters
   Editors: G. Benedek, T. P. Martin,
   and G. Pacchioni

7 Molecular Beam Epitaxy
   Fundamentals and Current Status
   By M. A. Herman and H. Sitter   2nd Edition

8 Physical Chemistry of, in and on Silicon
   By G. F. Cerofolini and L. Meda

9 Tritium and Helium-3 in Metals
   By R. Lässer

10 Computer Simulation
   of Ion-Solid Interactions
   By W. Eckstein

11 Mechanisms of High
   Temperature Superconductivity
   Editors: H. Kamimura and A. Oshiyama

12 Dislocation Dynamics and Plasticity
   By T. Suzuki, S. Takeuchi, and H. Yoshinaga

13 Semiconductor Silicon
   Materials Science and Technology
   Editors: G. Harbeke and M. J. Schulz

14 Graphite Intercalation Compounds I
   Structure and Dynamics
   Editors: H. Zabel and S. A. Solin

15 Crystal Chemistry of
   High-T_c Superconducting Copper Oxides
   By B. Raveau, C. Michel, M. Hervieu,
   and D. Groult

16 Hydrogen in Semiconductors
   By S. J. Pearton, M. Stavola,
   and J. W. Corbett

17 Ordering at Surfaces and Interfaces
   Editors: A. Yoshimori, T. Shinjo,
   and H. Watanabe

18 Graphite Intercalation Compounds II
   Editors: S. A. Solin and H. Zabel

19 Laser-Assisted Microtechnology
   By S. M. Metev and V. P. Veiko
   2nd Edition

20 Microcluster Physics
   By S. Sugano and H. Koizumi
   2nd Edition

21 The Metal-Hydrogen System
   By Y. Fukai

22 Ion Implantation in Diamond,
   Graphite and Related Materials
   By M. S. Dresselhaus and R. Kalish

23 The Real Structure
   of High-T_c Superconductors
   Editor: V. Sh. Shekhtman

24 Metal Impurities
   in Silicon-Device Fabrication
   By K. Graff   2nd Edition

25 Optical Properties of Metal Clusters
   By U. Kreibig and M. Vollmer

26 Gas Source Molecular Beam Epitaxy
   Growth and Properties of Phosphorus
   Containing III-V Heterostructures
   By M. B. Panish and H. Temkin

* The 2nd edition is available as a textbook with the title: *Laser Processing and Chemistry*

**PROPERTY OF
ERAU PRESCOTT
LIBRARY**